TRANSFER OF TECHNOLOGY

TRANSFER OF TECHNOLOGY

U.S. Multinationals and Eastern Europe

Marilyn L. Liebrenz

PRAEGER

PRAEGER SPECIAL STUDIES • PRAEGER SCIENTIFIC

Library of Congress Cataloging in Publication Data

Liebrenz, Marilyn L.
Transfer of technology.

Bibliography: p.
1. Corporations, American—Europe, Eastern.
2. Technology transfer—Europe, Eastern.
3. Technology transfer—United States.
I. Title.
HD2844.L53 338.91′73′047 81-17908
ISBN 0-03-061344-2 AACR2

Published in 1982 by Praeger Publishers
CBS Educational and Professional Publishing
a Division of CBS Inc.
521 Fifth Avenue, New York, New York 10175 U.S.A.

23456789 052 987654321

Printed in the United States of America

For continuing confidence and patience,

this book is dedicated to my parents,

Emma and Allan F. Liebrenz,

and to my brother and his wife,

Allan Lowell and Janet Liebrenz

FOREWORD

Eric W. Hayden

The transfer of technology to socialist countries involves some serious issues for U.S. policy makers, particularly when such transfers permit the recipients to reallocate resources so as to focus on developing other productive capabilities. Not to underestimate the very real importance of such potential national security issues, however, it must also be stressed that long-term enterprise-to-enterprise relationships of the type fostered by technology transfer arrangements can be beneficial to the United States. One example is the jobs provided for those who produce the components and equipment sold in conjunction with the transfer of the particular technology. Equally important are the exports to be realized, since they help reduce what has become in recent years a permanently negative U. S. trade balance. For a variety of reasons, these and related gains have become increasingly accepted, and even appreciated, by many of those who once argued most vociferously against U. S. firms transferring their technology abroad.

But besides the macro benefits accruing to the United States from such activity are the equally significant gains realizable at the micro level by those firms venturing to sell their technology abroad. It is in analyzing these benefits, particularly as they apply to firms dealing with Eastern European nations, that Marilyn Liebrenz has made a meaningful contribution to the dynamic field of international business.

First, she has provided a timely and much appreciated look at several technology transfer cases that I had initially examined several years ago at their respective inception stages. In the intervening years, events have impacted the cases in a variety of ways. Whatever changes time may have brought, however, Dr. Liebrenz shows that the fundamental premise of my earlier work has withstood the test of time, namely, that nonequity tie-ups with Eastern European partners are both workable and sustainable. This conclusion not only provides some satisfaction to me personally but, far more important, confirms the faith of those corporate pioneers who braved many initial doubts in the early 1970s to expand the nature and extent of their respective firm's international activities. Liebrenz's findings are that much more meaningful, since she also has added new case studies, thereby broadening the range of successful technology transfers to include not only Poland and Romania but also Hungary and Yugoslavia.

A second major contribution of this book is that it provides an all-encompassing list of factors that any would-be participant in a technology transfer arrangement would be well advised to consider, regardless of the extent of prior experience in transferring technology and regardless of whether the purchaser is a socialist- or capitalist-oriented producer. Some of the descriptive as well as analytical factors Dr. Liebrenz develops are worth mentioning here, since they illustrate how thorough is her understanding and coverage of the major issues of technology transfer. They include such technologically related factors as the product/process involved, skills and training tasks, and language barriers; supportive capabilities, like plant construction, layout, and supply; environmental considerations ranging from cultural interfaces to such fundamental logistical matters as expatriate housing; leverage and control, even in the absence of equity by the Western firm; and the rationale for, and benefits from, the transfer.

Besides the importance of such factors in ensuring a successful technology transfer arrangement, there is another aspect that emerges from Dr. Liebrenz's investigation. Obviously of significance in any sale—whether of a product, a factory, or a technology—is the quality of the item being sold. Nevertheless, as the case studies and analysis of this work make so very clear, there is much more involved in a successful transfer arrangement than simply the technology package per se being offered by the successful bidder. Selling technology requires developing a long-term relationship between supplier and purchaser, which means that the successful vendor needs to develop and practice business methodologies that are frequently judged to be more characteristic of such formidable global business competitors as the Japanese—for example, long-range planning, flexibility and adaptability, commitment to follow-up, and interpersonal relationships. While it has become topical to criticize the shortcomings of U.S. business in these areas, it is certainly encouraging to point to the counterevidence offered by those successful firms documented in this study.

PREFACE

Technology transfer remains a controversial issue in U.S. foreign and international business policy. Despite the strife in Poland in the early 1980s, trade between the East and the West climbed from \$15 billion in 1970 to more than \$106 billion in 1980, as reported by the U.S. Department of Commerce. In fact, the United States led all Western countries in agricultural exports to Communist countries. Proponents stress the benefits to the U.S. balance of trade through sales of U.S. technology abroad; opponents continue to administer warnings based on the stance of the United States with regard to its diminishing technological lead in many product areas. Both sides admit to concerns regarding the creation of potential competitors through such technology sales. The following case studies contribute, by reflecting actual experiences, some answers to such questioning.

This book builds upon the prior study of East-West technology transfers researched by Eric W. Hayden in the early to mid-1970s. His extensive documentation of selected transfers provided a framework for categorization of technology sales not only applicable to Eastern Europe but also internationally. This work extends the foundation provided by Dr. Hayden, utilizing his research as a base for identifying corporate expectations. It follows the descriptive areas distinguished by his studies and looks at a number of the same transfers after the passage of several years.

The conclusions in this book comprise three levels. The first level, perhaps the most beneficial for those companies and countries engaged in technology transfers, describes the collective experiences of the eight transfers. The selected research format, which compares corporate expectations against actual transfer results, highlights discrepancies. Despite the relatively specific nature of these case studies, the collected information is widely diverse: basic technology to extremely intricate and complex information; transfers completed within five years to those lasting twenty years or longer; low-level, small component manufacturing to complete, heavy equipment machinery production. Even the Eastern European countries provide a spectrum within the framework of communist ideology: Yugoslavia, Poland, Hungary, and Romania. The location is Eastern Europe, but the scenario might reflect transfers in numerous parts of the globe.

Level two is more analytical, identifying certain characteristics of technology transfers that tend to be problem areas. These results, drawn from the transfer descriptions, should help corporations to structure future technology sales.

The problem/opportunity approach and the maintenance of leverage or control are among the major issues addressed in the third level throughout this research. Future competitive threats and the maintenance of leverage or control are addressed on this level. The final answers to these question areas are dependent on the future, but the current positions of the corporations involved in these technology sales illustrate how they dealt with these issues.

The completion of the research comprising this book was aided to a great extent by a number of individuals from business, government, and academia. Their help, information, and support were essential. This study would not have been possible without the extensive cooperation of the selected corporations: Bendix, Clark Equipment, General Tire International, Honeywell, International Harvester, and Westinghouse. These companies graciously provided the staff contact and corporate documentation necessary to detail these transfers. To all involved, my grateful appreciation is extended.

While it is not possible to thank everyone by name, I would like to mention several individuals whose help was of particular value at various stages during the creation of this book. Eric W. Hayden has been mentioned earlier, and a large note of thanks is directed to him. A major portion of the groundwork for this study was laid by his extensive research. Without his professional support and personal encouragement, this work would not have been successful. Thanks also goes to special individuals within the academic community, including Donald A. Taylor, Donald S. Henley, John L. Hazard, and Larry W. Foster who helped from the formation of this study to the final framework that resulted from the analyses. Numerous professors at Michigan State University helped to provide a research foundation. Special thanks goes to, among others, Dr. William Lazer and Dr. Gil Harrell. Clark E. Dehaven provided constant encouragement throughout the entire work. Bringing his knowledge of the subject, the insight of John K. Ryans, Jr., greatly helped in the composite of the final results.

Finally, I would like to pay tribute to my ever-patient typist, Nancy Hodge; to my skillful graphics artist, Michael McClain; and to Marie Iliff who literally nurtured me throughout this project. To these individuals and to my family and other friends, I send grateful appreciation. I hope that I will someday be able to return to others some measure of the support and encouragement given to me.

CONTENTS

LIST OF TABLES AND FIGURES

TRANSFER
OF
TECHNOLOGY

1

PERSPECTIVE ON NONEQUITY TECHNOLOGY TRANSFER TO EASTERN EUROPE

INTRODUCTION

Multinational corporations (MNCs) have utilized different international linkages in their global approaches. Some linkages involved a form of direct investment, a means by which an MNC could gain continuity and, ideally, a high degree of control. Another linkage has been evolving, however, that is viewed as less traditional. This is the sale of technology that does not include an equity investment on the part of the MNC in the project. Furthermore, unlike several other linkages, it is never expected to evolve, in most cases, into a direct investment or a joint venture.

This type of linkage involves the transfer of the corporation's technology and/or managerial skills in return for payment from the foreign party. This knowledge sale differs from a typical product sale in several respects. First, the foreign party may be a private company, but it may also be a foreign government. Second, the transaction is viewed as a onetime sale and purchase although the time involved in the transaction may encompass a number of years. Third, the control that the MNC has over the arrangement exists but usually differs from that held in an equity investment. A high degree of control over the operation may not be viewed as desirable or even necessary.

This transfer of technology also differs from more customary licensing arrangements as well. The technology transferred via sales usually involves a license arrangement, but the qualitative and often the quantitative aspects of the technology are generally far more extensive than prior licensing concessions. The magnitude of the projects—which often involve some construction and equipping—tends to be more typical of arrangements usually accompanying direct investments.

Justification for this nontraditional transfer approach appears to rest more on necessity than choice. Fears concerning market saturation in developed countries are growing in numerous industries. Trading opportunities in countries that permit direct investment by the MNCs have been diminishing in recent years. In many cases new market opportunities that open up are in the socialist world sphere or with developing Third World nations. Many of these opportunities, however, include restrictions on MNCs entering these markets.

In 1974 researchers noted that MNCs were finding it increasingly practical to stress their technical and marketing skills without trying to obtain ownership or control of foreign property.[1] This tactic was used particularly in dealings between U.S. companies and the USSR. The Soviets desired U.S. technology and marketing expertise but would not allow foreign ownership or control. Developing nations, too, especially those with reserves of key raw materials, were demanding more of a say—and often a controlling interest—in MNCs operating within their boundaries.

Raymond Vernon noted these tendencies in developing nations. The concluding chapter from his book Storm over the Multinationals states:

> One clear line of policy in many developing countries has been to provide support to local public and private enterprises in their efforts to find untied sources of foreign capital and technology, and to develop their own channels of local and foreign markets. With that objective in mind, governments have imposed special taxes on the income of foreign-owned subsidiaries, ceilings on license fees paid by local licensees to foreign licensors, limitations on foreigners' use of local credits, requirements for the sale of equity to local interests, and so on.[2]

In an analysis of another potential marketplace now opening to the West, the People's Republic of China (PRC), John K. Ryans, Jr., wrote that "the PRC . . . has been attempting to achieve a major increase in agriculture and develop its industrial capability through its own resources and initiative."[3] The result is that, for one or several reasons, corporations are applying the alternative of nonequity transfer of technology and/or management instead of using the linkage of direct investment in arrangements beyond the usual licensing operations.

This book compares the perceived outcomes of MNCs that have entered into nonequity transfers of technology and/or management with their initial expectations. Incentives behind the usage of the nonequity transfer will be examined. Successful technology transfers may take many forms; the attainment of every expectation in a nonequity transfer may not be a requirement in terming the venture a success from the viewpoint of the MNC. However, if future nonequity projects are to be a viable possibility, MNCs will have to achieve certain goals or a minimum level of satisfaction of the expectations that were present when the agreement was arranged. This study aids in the identification of problems that may arise and offers possibilities for avoiding or circumventing those problems, thus enabling more MNCs to avail themselves of the economic opportunities present via technology transfers.

The material presented here has been organized into five chapters. Separate corporate descriptions regarding expectations and outcomes will be found in Chapters 3 and 4.

Chapter 1: Presentation of a statement of the situation, including background on nonequity agreements, technology transfers, and an overview of Eastern Europe through a review of the literature;

Chapter 2: Statement of methodology and areas of concentration as derived from the literature and preliminary corporate interviewers; description of question areas used in conducting the study;

Chapter 3: Description and analysis of corporate expectations in nonequity technology transfers to Eastern Europe;

Chapter 4: Description and analysis of corporate outcomes of nonequity technology transfers to Eastern Europe;

Chapter 5: Summation of results with conclusions, limitations, and recommendations for further study;

Appendix A: Listing of corporations, countries, and products involved with the eight agreements considered in this study;

Appendix B: Questionnaire outline used as basis for personal interviews.

BACKGROUND

Development and Acceptance of Technology Transfer

Technology appears to be one of the most important of U.S. exports. Royalties from the transfer of technology amounted to $2.95 billion in 1977.[4] However, much of the research that has been done in the area of technology transfers has focused on the appropriateness of such transfers as to local product adaption or on the creation of a code of conduct covering the transfers. Instead of examining the transfer arrangements, studies usually dealt with the desirability of technology transfers, concentrating mainly on the feasibility of such linkages.

Technology transfer may be broadly interpreted to include all items reflecting a firm's knowledge of a certain product or process that are passed from that firm to other enterprises. An accepted meaning for technology transfer now indicates that it is the transfer of the ability to reproduce an item or a part of the item. One source stresses the fact that transferring such ability requires a long-term, know-how commitment on the part of the seller via selected personnel who are both familiar with the product being produced and able to adapt to foreign conditions. It is also noted that the long-term commitment often involves updating obligations during the time of the agreement, and it sometimes includes a buy-back arrangement.[5]

In the early 1970s the basic structures used in technology transfers were licensing agreements, turn-key arrangements, coproduction agreements, and operation or management contracts.[6]

Examples of technologies "for sale" might include those related to[7]

Aiding in the production of a product
Providing services to other firms
Aiding product design
Enhancing product value
Branching out to nonprofit areas
Strictly selling know-how

Such technology sales are sometimes overlooked by some corporations, possibly because of increased difficulties associated with marketing a relatively intangible product (such as personnel management training), or because such sales opportunities are not recognized in the first place.[8]

A broadened view of technology, extending beyond the realm of a particular tangible product, was emphasized by Eugene Lang who noted that technology transfer may "consist of patents or trademarks, an ability to do something uniquely well, a good product design, special application techniques, or anything that is sufficiently distinctive to permit the individual manufacturer to be a strong competitor in a particular market."[9]

An international survey of chief executives was made by James R. Basche, Jr., and Michael G. Duerr for the Conference Board in 1975. This survey considered all possible avenues for technology transfers, including investment ventures, and highlighted the opportunities available to corporations and the various reasons for seeking such technology sales.[10]

Another comprehensive study of technology that emerged in 1975 was submitted by Harvey W. Wallender to the United Nations Economic and Social Council at the Seminar on the Management of the Transfer of Technology within Industrial Cooperation.[11] One of the more important points of his report was the detailed breakdown of technological transfer options. This breakdown highlighted the level to which the analysis of technology transfer had risen by the mid-1970s over the prior two decades.

Arguments against Technology Transfers

Arguments against the transfer of technologies are numerous, if not conclusive. Basic criticisms appear to fall into four categories: exportation of jobs, creation of competition, national security issues, and maintenance of control.

Exportation of Jobs

For Americans, job exportation has been an argument against several forms of international linkages, including technology transfer. Hawkins and Finn noted that while the United States has traditionally adopted a liberal policy toward multinational corporations, the movement toward increasing federal regulation of MNCs in order to restrict foreign investments and technology transfers has been supported by U.S. labor.[12] Winpisinger seconded this view by pointing out that the technological innovations that are produced in the United States and then transferred to other nations cause many U.S. workers to be unemployed.[13]

It has been validated, however, that technology transfers, at least in the short run, at times help to increase jobs and may even be vital to the continued growth of industry.[14] Harvey and Kerin espoused this view, stating that MNCs have had a definite positive effect on domestic employment—they have not decreased domestic export-related jobs through worldwide distribution of technology.[15]

Creation of Competition

Creation of potential competition appears to be another chief argument against technology transfer. One rebuttal against the competition argument has been that, in many industries, the rapid pace of development is such that the transferred technology is often outdated in a short period of time. For example, computers comprise one field in which innovations usually keep the state of the art ahead of the technology being transferred. However, as reported in Computerworld, the U.S. computer industry has begun to experience diminishing markets in Europe and Japan.[16] The uncontrolled transfer of U.S. technology to other countries has been blamed for the stimulation of competition that could threaten the U.S. industry. U.S. export controls have not been at fault, however; the problem apparently stems in large part from the fact that countries receiving U.S. technology have been passing it along to additional countries.[17]

One common complaint involves the export of U.S. technology and, indirectly, U.S. jobs. A point in favor of curbing the sale of U.S. technological innovations abroad was made by Herbert Meyer, who wrote that the export of technology will only hurt some other U.S. firm because many nations are using this very knowledge to undersell U.S. firms in world markets. Prohibiting the sale of U.S. technology, comments Meyer, would encourage export purchases, providing more royalties for U.S. firms while maintaining that competitive edge.[18]

Marketing the products of technology previously gained via a transfer appears to be a logical development. Referring to a particular case, one author noted, however, that technology purchased from France by Eastern Europe was probably being used to manufacture products that were later being shipped back to compete with other products in Western markets. This appears possible, since Eastern European countries are eager to export goods to improve their hard currency shortage.[19] It is also not illegal, unless specifically forbidden by the technology transfer agreement.

Lang dismisses the arguments against competition by stating that the question is not really valid. He presents the following four points favoring technology transfer:

> First, know-how can be dealt with as property value, protected by ground rules in most countries. Second, contractual techniques can limit its transfer to others. Third, those concerned about revealing their know-how are often kidding themselves. A competitor, given enough time, money and talent, can acquire a similar body of knowledge. Fourth, know-how is a wasting asset—it becomes obsolete as technology advances and markets change.[20]

One aspect of technology transfer supporting its international sale and distribution is that the sale of technology abroad has encouraged firms to invest more capital in research and development —capital that is generated by current international technology sales. Studies are finding that U.S. corporations tend to export new products from foreign sources much earlier than in the past—within five years now. It is also pointed out that greater secrecy is maintained regarding processes than products these days.

Overseas operations are not just on the receptive end of such transfers. These operations also provide some of the revenues as well as the inspiration for development. One source states that approximately 30 percent of the financing for research and development stems from overseas plants, and some development of the global approach to innovations has been conducted abroad as well.[21]

Competition as a whole is not an issue for all companies. One U.S. company, for example, appears more concerned about patent protection and cash flow issues than about foreign competition. The position of Texas Instruments, as stated by its general patent counsel Harold Levine, is that "no transfer of high technology should take place unless it (is) in a manner which provide(s) a sufficient flow of funds into the U.S." The company notes that it feels only government intervention can secure patent protection.[22]

Clarification of technology issues with regard to competition appears to require a breaking away from generalities and a concentration on more specific issues. For example, a recent article by John Kiser noted that discussions on the "technology gap" refer to averages, which place the United States in a superior position. In reality, however, Kiser points out that technology from the USSR is being used routinely by U.S. companies. He urges that a deeper analysis be made to uncover those areas where each country is technologically superior.[23]

National Security Issues

The question of compromised national security is the third major contention against technology transfers, although the global activities of MNCs as a whole have also been a focal point for such considerations.[24] Root points out that the Export Administration Act is aimed at restricting U.S. exports to politically sensitive areas.[25] Technology transfer export laws involving defense controls have been particularly applied to technology sales to the Soviet bloc.

The position of the Department of Defense is that the transfer of some types of technology is tantamount to providing a country with the means to conduct more sophisticated warfare against the United States.[26] The rationale underlying export restrictions comes from the belief that high technology is vital to the strategic security of the United States.[27]

The United States has been described as having the most stringent controls over the export of high-technology products of any Western country.[28] Those in contention with current export controls take the position that the lack of criterion and conflicting interpretations of export regulations, combined with the lack of distinctions between military and nonmilitary technology, make it extremely difficult for U.S. industry to assume a responsible position in such technology transfers.[29] Those concerned with the unfavorable balance of trade currently held by the United States note that restrictions on the export of products with potential strategic applications (of either the product or the know-how) to Communist countries are seriously curtailing U.S. sales abroad. They are also restricting companies competitively through bureaucratic delays and licensing procedures.[30]

Maintenance of Control

Direct investment traditionally provided both partners with some element of control, primarily control usually residing with the majority partner. However, at least some corporations are observing that even though an MNC possesses a majority interest in a company, local governments in some countries exercise de facto control over the company anyway.[31] Technology transfer offers many of the benefits ascribed to international business without the risk of losing financial investments: establishment of new markets, increase in corporate reputation and product demand, and political and economic stability assessments.

Technology transfer sales appear highly viable in the area of natural resource development. While the U.S. company may hold a high degree of control at the start of a direct investment project of this type, readjustment of the initial contract becomes highly probable once the project is in operation. The sale of management services without the accompanying equity investment can provide an income that is unlikely to be cancelled.[32] These arrangements, sometimes called service agreements, appear to be on the rise and are seen as up-and-coming opportunities.[33] See Figure 1.1 for a diagram of the four problems here.

FIGURE 1.1

Problems Accompanying Technology Transfer

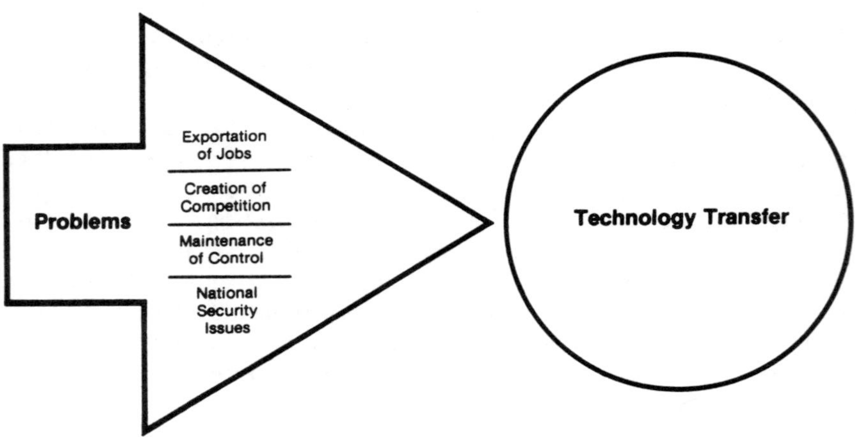

Selection of Technology Transfer Linkage

Viable Alternative

Would an MNC select technology transfer as the most advantageous linkage over other options, such as 100 percent investment or joint ventures? Such a transfer is often a viable alternative for the financially strapped corporation, since "governments all over the world are attempting to get a bigger share of the profits, jobs, markets, and technical and managerial skills that multinational companies create or control."[34] As these MNC-controlled "benefits" are desired without accompanying ownership strings or import fees, technology transfers are often one of the few feasible access routes to some markets. The transfer provides financial remuneration to the MNC and, in some instances, also makes alternative product production sources available for use by the MNC—without an equity commitment. Technology transfers do provide a middle-of-the-road approach that seems to provide more benefits with fewer detriments than many of the other linkages among current global options. David Horner of the Office of International Investment in the Bureau of International Commerce states that "licensing and management contracts offer a 'low profile' as contrasted to direct foreign equity investment . . . a 'middle way' of retaining foreign markets—a step beyond simple exporting, a step short of overseas production."[35]

Minimization of Political Risk

The business climate for direct investment ebbs and flows. In Latin America, for instance, foreign investment has been encouraged in some countries since the mid-1970s. It appears that sales of technology, including know-how, circumvent much of the risk associated with direct investment. One example of the sale of managerial expertise is the agreement that Exxon has with Venezuela to market, for a fee, Venezuela's crude oil and other products. This arrangement with Venezuela, which nationalized its oil industry in 1976, exemplifies a type of technology transfer that is growing in popularity around the world.[36] See Figure 1.2 for an overview of these issues.

Assessment and Improvement of Technology Transfer

One of the first studies to analyze technology transfers, including both technical and managerial knowledge, was published by J. S. Fforde in 1957. His inquiry was summarized as follows:

> An inquiry into such British experience as might be found relevant to the solution of the following general problem:

FIGURE 1.2

Advantages of Technology Transfer

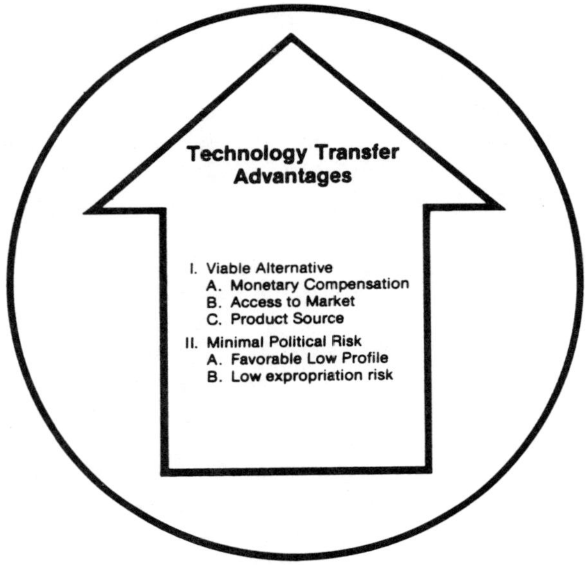

> if the underdeveloped countries should consider they require an element of foreign managerial and technical services for use in the operation of industrial, etc., enterprises locally promoted or supported, how may they best acquire and use such services on terms that are adequate to attract them in the correct quality and quantity but are also consistent with local social and political requirements.[37]

Fforde used the term managerial and technical services to include possession of a technical, commercial, financial, legal, administrative, or entrepreneurial bias. He distinguished four different routes in which this transfer occurred: (1) the individual route; (2) the joint venture (not necessarily implying equity); (3) the product-specialized foreign investment (only an outside firm involved, no local partner); and (4) the trader-industrialist (more a by-product of the bygone British era than a viable future route). Of these four transfer routes, the second and the fourth best reflect the type of nonequity transfer considered in this study.

It must be pointed out that, while Fforde described technology transfers, the desirability of such transfers or even the use of such transfers as an acceptable approach is another issue. Fforde noted that the situation of "local financial control [versus] foreign technical control" was more easily stated than performed.[38] He reported that this type of transfer was utilized by MNCs when the local situation appeared too politically risky, or when the market appeared initially lucrative but little hope was held for future growth. He did describe some MNCs that had switched from desiring an equity involvement to seeking short-run involvements. In these instances, they sold their skill and expertise instead of their product.[39]

Peter Gabriel, in the late 1960s, concentrated on analysis of the management contract (in contrast to Fforde, who included both managerial and technological skills in his study). The management contract, as defined by Gabriel, was an arrangement under which operational control of an enterprise or of one phase of an enterprise, which would otherwise be exercised by the board of directors or managers elected or appointed by its owners, is vested by contract in a separate enterprise that performs the necessary managerial functions in return for a fee.[40]

Gabriel noted several distinguishing traits between such a contract and several other types of contractual arrangements, such as technical services agreements, license agreements, engineering and construction agreements, and contracts for exploitation of mineral resources.

His research involved a detailed study of four different situations that utilized management contracts, describing the means by which the contract was established, general terms, and a final evaluation based on the operation of the contract. He concluded that management contracts could make significant contributions toward increasing the efficiency of public enterprises in developing countries.[41]

The use of management contracts was also studied by British researcher Richard Ellison, who did an indepth investigation of the management contract activities of 16 large MNCs. He noted that "for many companies, the management contract, together with licensing and technology agreements, appears to be the only way to maintain or build their interests in many parts of the world."[42]

Furthermore, he found that while approximately half of the companies he studied used management contracts for "defensive" reasons, particularly protection of an investment that had been expropriated by a local government, the other half used them "aggressively" in actively looking for situations in which they were able to sell their services. Ellison proposed seven suggestions toward a more successful operation utilizing this approach.

1. A company must work hard at changing its traditional attitudes
2. Rapid training of local nationals is essential
3. Control must be shared
4. Equity participation should usually be avoided

5. Fees dependent on profits can be dangerous (better on sales volume)
6. Some internal organizational changes are needed
7. If possible, use a short and simple contract document[43]

Management contracts apply to a number of industries, but one area that benefits greatly from the receipt of managerial expertise is that of mass merchandising. Clarence Mann, assistant general counsel for Sears, Roebuck and Company, provided an assessment of such expertise at the Seminar on International Issues in Transfer of Technology held in Paris in 1977. Mann's assessment of such service technology transfers was that the benefits extended far beyond just the immediate industry receiving the information, helping the entire country or region covered by the transfer.[44]

The identification of technology transfer as a feasible revenue source for companies elicited some studies toward improvement of such transfers. One of the few analyses to emerge, thus far, was conducted by H. J. Rath, who examined technology transfers between the United States, Canada, and the North Atlantic Treaty Organization (NATO) nations. His recommendations for technology transfers included the following four points:

> 1. For optimum success of technology transfers, monetary payments are preferable to a production (counterpurchase contract;
> 2. If counterpurchase is a prerequisite, the technology received should be listed in the agreement as the product sales vendor;
> 3. The technology receiver should be involved in the transfer from the beginning, in the planning stages;
> 4. The technology seller's personnel should act in an advisory capacity.[45]

The heightened attention being directed toward technology transfers should not be taken as a discovery of a new phenomenon. Technology, to some extent, has always been an underlying factor of license agreements. A relatively new aspect appearing within the past decade or two has been the increase in the amount of information provided as licensed technology, while corporate involvement on other fronts, such as direct investment or joint ventures, was conversely decreasing.

It may be more accurate, actually, to discuss the expanded growth of the information transferred via licensing than to concentrate solely upon technology transfers. The greatest bulk of technology transfers occurs through licensing. The impact of these international transfers has transformed the Licensing Executives Society in New

York, formed in 1964, into an international organization with chapters around the world.[46] The increasing use and importance of licensing was pointed out by a Business International publication, International Licensing: Opportunities and Challenges in Worldwide Technology Management, which appeared in 1977.[47] In addition to material applying to typical licensing conditions in selected markets, this publication lists alternative licensing strategies in global business strategies.

Licensing expertise offered by Eugene Lang includes the following cautionary notes:

1. Do not give licensee unqualified access to new developments;
2. Do not assume that use of the trademark can be properly licensed without restrictions against production of inferior products;
3. Do not warrant the validity of patents that you license, or blindly agree to attack alleged infringers;
4. Do not get involved with a licensee until you are absolutely certain he knows your particular line of business;
5. Do not set up license conditions that involve you in the management prerogatives of a licensee. Your job is to help, but not to make decisions for, the licensee;
6. Do not give up all control over the manufacture and marketing of your product abroad;
7. Do not assume that the specifications used in your organization can be followed abroad;
8. Do not think that mere agreement between you and the other party completes the deal—government approval is often required;
9. Do not assume that a large royalty or a minimum royalty is a guarantee if the licensee is not economically sound;
10. Do not assume that your know-how cannot be copied;
11. Do not think that just because an agreement has been reached and blueprints transmitted, profits will roll in—follow-through is essential;
12. Do not take anything for granted—ever.[48]

TRADE WITH CENTRALLY PLANNED ECONOMIES

Identification of Separate Trading Blocs

Centrally planned economies, with a combined population of more than 1.4 billion people, constitute a major potential market.[49] The Department of Commerce (DOC) reports that by 1977 "U.S. trade [imports and exports] with the centrally planned economies [including Eastern Europe, the USSR, and the People's Republic of China] totaled $3.8 billion, a dramatic increase from the pre-détente 1971 level of only $650 million."[50] The Department of Commerce also notes, however, that "while other industrialized nations of the Western world captured 25 percent of communist trade in 1977, the U.S. share was only 1.7 percent." Pointing out that the United States is responsible for 12 percent of world trade, the DOC notes that "there is apparently room for further U.S. penetration into the communist markets."[51]

This penetration did increase during the first nine months of 1979 as aggregate U.S. exports to the Communist economies totaled 39.7 percent above 1978 totals for the same nine-month period. As U.S. imports from these countries rose to 25 percent, a surplus balance of trade was achieved through U.S. exports to Eastern Europe (excluding the PRC and the USSR), which rose 20.4 percent.[52]

Trading relationships between the United States and the Soviet bloc have shifted since the mid-1940s. Gradually broadening and deepening from the end of World War II until the mid-1970s, the height of détente appeared to culminate with the 1972 Trade Agreement between the United States and the USSR. This was the "first attempt by the U.S. to establish a general bilateral trade agreement with a communist country," as noted by Irene Lange and James Elliott.[53] They point out, however, that the USSR never put the legislation into effect because the United States included aspects in its Trade Act of 1974 that were unacceptable to the USSR, particularly Section 402 (Jackson-Vanik Amendment) linking most-favored-nation (MFN) trading status to the right of free emigration, and Section 613, which placed a limit on the aggregate amount of credits available to the Soviet Union from the Export-Import Bank of the United States (Eximbank).[54]

One policy consideration that became clearer during the mid-1970s was that the socialist bloc was no longer being viewed as just one unit. The USSR, Eastern Europe, and the People's Republic of China were emerging as separate entities and were being treated as such. The USSR has not received most-favored-nation trading status from the United States, but Poland and Romania did receive this trading concession in 1974 and 1975, respectively, through the General Agreement on Tariffs and Trade (GATT). One indication that the United States intended to encourage closer East-West economic connections was evidenced in the creation of the Bureau of East-West Trade within the DOC. This bureau, established in 1972, was given dual functions: to "promote the orderly development of U.S. economic relationships with the centrally-planned economies of Eastern Europe and Asia and to continue the Department's responsibility for implementing and enforcing the nation's export control system."[55]

Countries included under the heading Eastern Europe vary from one source to another. The broadest interpretation encompasses all members of the Council for Mutual Economic Assistance (CMEA, formerly known as COMECON) plus Yugoslavia, which has observer status. Another common grouping includes the USSR but not Yugoslavia, and one of the narrowest refers just to Bulgaria, Czechoslovakia, the German Democratic Republic (GDR), Hungary, Poland, and Romania.

As of 1977, Eastern Europe, excluding the USSR and Yugoslavia, was listed as having a population of 108.1 million. Poland had the largest population at 34.9 million, followed by Romania with 21.8, the GDR with 16.8, Czechoslovakia with 15.1, Hungary with 10.7, and Bulgaria with 8.8 million people.[56]

Assessment of Western Involvement in Eastern Europe

The broadening of political acceptance of the socialist bloc in the early 1970s brought on a spate of agreements that, in turn, helped to generate additional studies of trade between the West and the East. Some of the analyses were directed toward increasing such trade.[57] The Columbia Journal of World Business in December 1973 devoted the majority of its entire volume toward East-West trade issues, including financing and marketing as well as political and economic assessments.[58]

Within a couple of years, however, the tone shifted. One of the predominant themes involved disappointments that resulted when earlier expectations did not materialize.[59] Another growing trend was the desire of the East to make inroads on trade with the West, as indicated in 1976 by Nikolai Patolichev, Soviet minister of foreign trade and others.[60]

A third theme continued a topic that had never completely died down from cold war times—security measures. The question "Is selling technology to the Soviets dangerous?" was never fully answered.[61] The danger alluded to stemmed primarily from the prospect of creating competition, but other dangers—to individual businessmen if not to the entire country—involved possible default and overborrowing.[62]

Studies concentrating on the market opportunity aspects of Eastern Europe, apart from the Soviet Union, came into prominence in the 1970s. In most instances, "Eastern Europe" included all countries that were members of the CMEA, usually excluding Yugoslavia which had only observer status. Observations were made not only by representatives of the West. Comparisons of the impressions of Eastern Europeans regarding MNC corporate officials working on projects in Eastern Europe were also reported.[63]

Eastern Europe began to be identified as an area to be studied, either alone or in connection with multinational corporations. One study by G. Peter Lauter and Paul M. Dickie laid the groundwork for future research by pointing out the value of such analyses.[64]

One of the earliest guides to business involvement with Eastern Europe was published by Business International S.A. in 1972. Excluding both Yugoslavia and the USSR, this lengthy publication provided both summary overviews of each country within its definition of Eastern Europe and chapter-long discussions of types of involvement, including "Organizing for East European Operations," "Developing a Marketing Strategy," "Negotiating the Contract," and other topics, ending with "Western Controls on Trade with the East."[65]

Manuals and monographs appeared by the late 1970s and offered guides and suggestions to aid executives in their negotiations and contracts with Eastern Europe. The Licensing Executives Society (U.S.A.) published a licensing guide to Eastern Europe (excluding the USSR) in 1977. Edited by Edward P. White, the volume addressed itself to a country-by-country breakdown of issues affecting licensing in each country.[66] In the same year, the Atlantic Council Committee on East-West Trade published a paperback designed to aid corporations in East-West trade.[67]

In 1978 an analysis entitled "East-West Trade Organizational Structures and Problems," by G. Peter Lauter, was presented at the Third United States-Romanian Seminar on Trade and Economic Cooperation in Bucharest.[68] The major thrust of the paper analyzed organizational structures that have been utilized by MNCs in their dealings with Eastern Europe—to overcome, in part, the criticisms of inflexibility and to enable the units to adapt to conditions within Eastern Europe. Lauter's basic premise was that "the large American multinational corporations, with a few notable exceptions (such as Control Data), have no comprehensive long-term East-West strategies and that, consequently, they have not developed the organizational arrangements necessary to fully exploit the various market opportunities in socialist economies. The limited integration of overall corporate activities and the difficulties of coordinating sub-divisional level operations result in cumbersome information processing and slowed-down decision-making."[69]

One case study was presented in Peter Lauter's paper on East-West trade organization, although the corporation's identity was disguised. Designed to point out organizational difficulties, the case study might also be used as a source of information as it also illustrates other problems facing corporations operating in Eastern Europe. However, while this study highlights problems that arise in an East-West trading transaction, it concentrates on one point in such a transaction rather than discusses changes or adaptions over time.[70]

A U.S. businessman involved in an East European project described his company's involvement with Romania. Thomas Fahey provided one of the few published observations written from the businessman's perspective with regard to nonequity transfers into Eastern

Europe. His observations regarding General Tire International's desire for such an involvement tended to emphasize the corporation's relatively small size, although accompanying factors were also apparent.[71]

The Department of Commerce also documented working conditions prevalent in Eastern Europe. It presented the results of a 51-company survey conducted in 1977 under the title Working Onsite in Eastern Europe and the Soviet Union.[72] Other DOC materials published in the latter half of the 1970s included such booklets as Joint Venture Agreements in Romania: Background for Implementation[73] and East-West Trade Financing: An Introductory Guide.[74] Produced through the Domestic and International Business Administration out of the Bureau of East-West Trade, these materials reflected the growing body of knowledge that had been amassed to promote trade with the Soviet bloc.

Certain aspects of MNC trade with Eastern Europe received more attention as circumstances warranted. For example, the countertrade issue was explored in several monographs and articles originating in the Department of Commerce.[75] Diagrams of countertrade steps have illustrated the steps involved in such transactions.[76]

It should not be interpreted that all materials were produced solely by organizations outside the Eastern European bloc. One monograph produced in West Berlin, "Guide to Trade and Cooperation with Eastern Europe," provided both an overview of information and additional suggestions for existing offices in Berlin from which more information was able to be obtained.[77]

TECHNOLOGY FLOWS BETWEEN THE WEST AND THE EAST

Analysis of Problems and Opportunities

In 1976 John P. Hardt and George D. Holliday made an assessment of the effects of Western technology on the socialist bloc. The impact was summed up in seven major points.

1. Western technology began to play a major role in the Soviet economy, starting with the Eighth Five-Year Plan (1966-70), carrying on into the ninth and tenth, respectively.
2. The selective policy of technology transfer and absorption appeared to have a significant aggregative effect on Soviet industrial production, far in excess of the modest value of imported machinery.
3. The effective transfer of technology was mainly in broader systems transfer, involving not only machinery and plants but also training, management, and foreign expertise or know-how.

4. The effective Western technology transfer, demanding domestic investment and adoption of Western systems, threatened to change the traditional role of the military and the party.

5. The rate of technology importation appeared limited by the Soviet ability to finance trade in hard currency.

6. Western technology was more critical to the economic modernization and consumer improvement plans of Eastern Europe. These countries were more flexible in their acceptance of Western commercial norms, though more constrained by balance of payment difficulties.

7. The People's Republic of China could engage in extended East-West commercial relations.[78]

Another survey published in 1976 presented opinions of corporate representatives actively engaged in technology transfers to the Soviet Union; it was highly optimistic of technology exchanges between the United States and the USSR. The six selected conclusions of the study are as follows:

1. U.S. executives with cooperation agreements with the USSR generally evaluated most aspects of the Soviet civilian system for generating new technology more favorably than those without such agreements.

2. The level of a company's technology did not appear to be an important factor contributing to the perception of Soviet technology.

3. Many of the businessmen involved with the USSR appeared to be well read and knowledgeable about the Soviet Union in general and highly literate about the areas of Soviet industry that directly related to their own business area.

4. The most well-read and best-informed businessmen seemed to have a reasonably high opinion of Soviet technology, with optimistic projections for valuable technology exchanges.

5. Many of the U.S. businessmen interviewed anticipated a minimum of risk regarding future Soviet competition.

6. Most of the businessmen interviewed saw arguments against U.S.-USSR trade as reflective of positions against international trade in general, not particularly against trade with Communist nations in particular.[79]

Not all studies involving technology transfers looked at the flow from the United States to the socialist bloc. One study, in 1977, considered the potential for technology transfers from the Soviet Union to the United States.[80] Surprisingly, 26 Soviet licenses were purchased by U.S. companies, twice as many as were purchased from the United States (13), and a contradiction to those who were unaware of this reversal of numbers. However, Kizer points out several factors that reduce the discrepancy to approximately 22 to 13. Furthermore, since he has established that the price paid for U.S. licenses is about ten times greater than the price paid for USSR licenses, the total number of licenses is misleading with regard to the size and magnitude of the trade. Kizer's personal reaction was that the "Soviet performance of selling technology abroad parallels their performance internally. That is, the reasons for a poor showing on the international market are related less to the availability of useful technology to sell and more to their (inadequate) system of incentives (poor), organizational structure, and scarcity of qualified (inclined toward a "selling" orientation) personnel for carrying out the job."[81] This does not reflect unfavorably on the USSR's technology per se, but it does present some barriers that might reduce a reverse technology flow.

Robert Schmidt takes a position supporting trade with the Soviet Union that holds that, through trade with the United States, the USSR will alter the nature of its society, enhancing opportunities for more freedom and compatibility with U.S. society. Schmidt sees economic survival for the United States as based on its ability to trade, and he points out that, though the United States has products and technology as its trading resources, without exchange these resources will stagnate.[82]

Interestingly, another analysis made on the impact of imported motor vehicle technology on the Soviet Union derived the following conclusion: "It is apparent that Soviet leaders are strenuously urging domestic industry to benefit from imported foreign technology. While significant diffusion undoubtedly has occurred, there are signs of Soviet dissatisfaction over the rate and extent of diffusion. Soviet industry in general has not been transformed through infusion of Western technology and there is little sign that industry will be revolutionized in the near future."[83]

One problem impacting on U.S. trade to the socialist bloc concerns export restrictions. Bureaucratic delays result from a case-by-case analysis of each product, and at least one author feels that the United States needs a coherent national policy of technology transfer based on three objectives: (1) improved control over militarily significant technologies; (2) minimization of the impact of controls on commercial trade; and (3) encouragement of continued development of new technologies.[84]

An accompanying problem related to export controls, writes another researcher, William Norris, is that there is a lack of positive initiative. Export control guidelines are based largely on U.S. assessment of Communist technology.[85]

The opinion in some areas is that the United States will, in the future, continue to be the world's technological leader and that the real threat to world peace lies in the quantity, not the quality, of Soviet weapons and in Soviet willingness to use these weapons rationally.[86] A 1978 study by the Brookings Institution concluded that no military, political, or economic reasons existed for denying exports of sophisticated data processing equipment and technology to countries in the Eastern bloc.[87] The rationale behind this position was that levels of computer technology in the United States and Communist countries were perceived as being relatively equal. Furthermore, such technological exports were not viewed as substantially crucial to labor-intensive economies.

Many military technologies also serve important peaceful purposes, though some uncertainty exists as to the end uses of such technologies.[88] The other side of the coin, however, is that products with civilian uses also have military significance. Despite projections that the United States and the USSR have achieved fairly equal levels of computer technology, at least one source supported curtailment of the Control Data computer sale to the Soviet Union. In this case, objections were based on the fact that the computer possessed at least 40 times the capability of any computer then held by the USSR; questions also existed as to the ultimate use for which the machine was intended.[89]

The basic value of export controls in general has also been questioned. The spread of critical technology may be only retarded, not stopped, by such restrictions, since this technology reaches Communist countries through nations friendly with both the Western and Eastern spheres.[90] Assessment of the security issue from a chronological perspective was provided by Samuel P. Huntington in the article "Trade, Technology, and Leverage: Economic Diplomacy," which appeared in 1978.[91] According to Huntington, economic relations between the United States and the Soviet bloc were at their lowest during the cold war of the 1950s, as exemplified by the creation of the Coordinating Committee on Export Controls (COCOM). The pendulum swung the other way, however, with the opposite position appearing in the early 1970s. The argument then propounded was that "American economic dealings with the Soviet Union should, so far as possible, resemble those with any other society."[92] Regardless of political consequences, Huntington noted that the results of this détente have benefited the Soviet Union more than the United States or the West in general, but he points out that the Soviets are also now facing other problems resulting from this advanced economic development.[93]

Studies of Technology Transfer to Eastern Europe

Many of the analyses and studies that have looked at U.S. business involvement in Eastern Europe have included technology transfers in one form or another. Three studies, however, specifically addressed the issue of technology transfer to Eastern Europe. Hayden's study in the mid-1970s chronicled a number of nonequity transfers between U.S.-based MNCs and countries in Eastern Europe. Because of the restrictions against equity involvement for most countries in this region, few of the connections involved equity. His study included management contracts as part of the whole area of technology transfers. Technology was "the quantum of knowledge by which such inputs as patent rights, scientific principles and R&D are translated into production of marketable materials, components, and end products. This quantum comprised two parts: the engineering documentation and the manufacturing techniques, with the latter being the human implementation of the written instructions contained in the former."[94] Hayden's objectives were to describe the types of technology transfers being supplied to Eastern Europe by U.S.-based multinationals; to provide a framework of categories of such transfers; and to sort out, as far as possible, the characteristics and advantages and disadvantages associated with each transfer vehicle. His intention was to put together as complete a description as possible at that point in time. His particular analysis detailed the types of arrangements made by firms, noting that there were three stages of involvement in this nontraditional, technology area: (1) technical assistance, (2) technical assistance plus trademark rights, (3) technical assistance, trademark rights, and repurchase commitments.[95]

The second major study was a massive one conducted prior to 1976 under a contract from the Department of Commerce through the Bureau of East-West Trade. Much of the data was collected by Paul Marer and Joseph C. Miller through the International Business Research Institute of the School of Business, Indiana University, with many of the corporate interviews conducted by John B. Holt. Unfortunately, this comprehensive report, East-West Industrial Cooperation: The U.S. Perspective, remains as yet unpublished by the DOC, but Marer and Miller provided a summary of some of the findings in their articles in the Journal of International Business Studies in 1977.[96] Their work involved all CMEA countries (excluding Yugoslavia) and encompassed industrial cooperation agreements (IDCs), including those of an equity nature as permitted by Romania and Hungary. Their initial mail questionnaire was sent to 1068 corporations with some type of CMEA involvement, with follow-up interviews conducted either in person or by telephone. Marer and Miller used an eight-category descriptive listing to distinguish types of IDCs: scientific technical agreements, know-how, licensing, turn-key, subcontracting, coproduction, joint ventures, and reverse licensing (the U.S. firm imports a license). The scientific technical agreements were not actual contractual projects but areas of exploration that, hopefully, would lead to a project.[97]

The third study, published by Business International S.A. in November 1977, was titled "Corporate Strategy, Planning, Organization and Personnel Practices for Eastern Europe."[98] The report included the USSR as well as Yugoslavia and considered corporations from Western Europe, North America, and Japan. This study was unique in that it "represented the first attempt to systematically investigate the experience of a large number of Western companies in a variety of industries which had been active in Eastern Europe for many years." The objectives of Business International (BI) were to describe what had been assumed were well-defined sets of strategies and tactics used by corporations involved in Eastern European regions. It became apparent to the researchers, however, that such corporate policies had not been planned out ahead of time by the companies, but were adopted ad hoc. BI therefore provided descriptions of the involvements without statements of corporate policies. One concern cutting into the value of the study was that "in view of the need to maintain good relations with their Eastern European business partners, many companies were reluctant to discuss in depth the many problems they continually face and how they overcome them."[99]

FRAMEWORK FOR ASSESSMENT OF TECHNOLOGY TRANSFER TO EASTERN EUROPE

Identification of Crucial Issues

Technology transfer has been found to be a workable global linkage for multinational corporations. The major criticisms against such transfers appear to fall into four categories: exportation of jobs, creation of competition, threats to national security, and maintenance of control. Criteria weighed in the selection of a technology transfer over other linkages include monetary compensation to be gained; access to an otherwise relatively inaccessible market; the possibility of establishing a nonequity product production source; and the minimization of political risk, especially regarding expropriation.

The exportation of jobs is an issue that is dependent upon the situation. Eastern Europe—with the exception of Romania, Hungary, and Yugoslavia—does not permit joint ventures of an equity nature with firms based in capitalistic-centered economies. Those joint

FIGURE 1.3

Ongoing Competition and Control Concerns

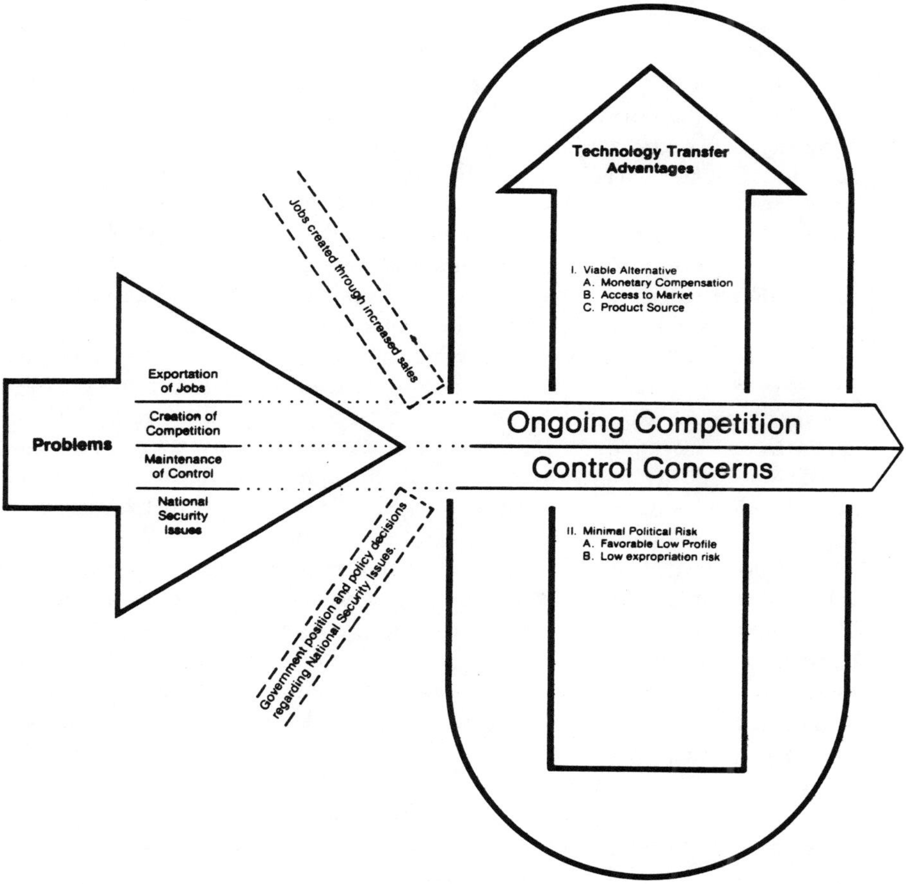

ventures that are permitted are thoroughly scrutinized. The shortage of hard currency throughout Eastern Europe makes exporting to that area a virtual impossibility without alternative financing measures. Thus, technology transfer, particularly with buy-back or countertrade measures, is often one of the only means available for entering the Eastern European market. In such instances, job exportation is a moot issue, since alternative sales are not feasible.

National security issues, particularly the use of U.S. technology to gain military superiority, are a chief consideration in dealing with communistically controlled economies. However, the decision as to what technology may or may not be transferred is not so much an issue for the individual corporation as for the government agencies responsible for ruling on such exports.

Thus, the decision facing the multinational corporation centers around the two remaining issues: the creation of competition and the maintenance of control. Competition is more or less a consideration of the future. Firms setting out to transfer a certain body of techniques, information, know-how, and processes face the possibility that in direct proportion to the success with which they are able to complete such a transfer, they create a competitor with powers potentially equal to their own, at least as of a certain point in time.

The control issue is also vital. Without equity, either on a majority or even minority basis, the corporation must find its leverage in other areas. Since the corporation's strongest position is at the initiation of the agreement, or even in the negotiating stage prior to initiation, the maintenance of control throughout the transfer is an issue—if not a requirement—during the entire operation of the agreement (see Figure 1.3).

One advantage of technology transfers is the option of increased trade when other forms of access are impossible or undesirable. Technology transfer also offers the MNC a viable additional source of components or completed products, as well as access to additional knowledge through contact with other firms and individuals. Furthermore, it does so with a minimum of political risk, in comparison to direct investment alternatives.

Expectations versus Outcomes of Technology Transfer to Eastern Europe

Research done on nonequity transfers has dealt with descriptions of the transfer, usually as related to a particular aspect. Suggestions for successful nonequity transfers, based on the researcher's observations, have been included in some of the studies, but these have not been the primary focus of the research. The only study that examined expectations versus actual outcomes was done by Gabriel, but this dealt specifically with the management contract approach and not with the spectrum of possible transfers. No study appears to have looked specifically at expectations versus outcomes for nonequity transfers by examining particular "real-life" transfers over a period of time. The possible use of such nonequity technology transfers warrants such a study; improving such technology sales reinforces this need. The scarcity of research in the area of nonequity transfers, combined with the lack of experience on the part of most MNCs in this area, provides a need for the collection of as much experience and predictive analyses as possible.

The following study concentrates on an assessment of eight separate technology transfer arrangements to selected countries in Eastern Europe, with the intention of examining the techniques used by the MNCs involved in those transfers. Factors are identified to help ensure the success of future technology transfers. The creation of competition at the conclusion of the venture and the maintenance of negotiating leverage (that is, control over the transfer) throughout the time period of the agreement are elements that have been included in this analysis. The expectations that existed at the beginning of the transfer, though possibly subject to variation, will be compared to the outcomes at the agreement's conclusion. Observations and results, from which recommendations may be drawn for future technology transfers, will be of benefit to MNCs desiring to utilize this global linkage.

NOTES

1. W. David Gibson and John Price, "A New Ball Game for Multinationals in the Offing," Chemical Week 18 (December 1974): 29.

2. Raymond Vernon, Storm over the Multinationals: The Real Issues (Cambridge, Mass.: Harvard University Press, 1977), p. 195.

3. John K. Ryans, Jr., "China: An Old Market Becomes the Newest Trade Opportunity," in China, The U.S.S.R. and Eastern Europe: A U.S. Trade Perspective, ed. John K. Ryans, Jr., Anant R. Negandi, and James C. Baker (Kent, Ohio: Kent State University Press, 1974), p. 15.

4. Herbert E. Meyer, "Those Worrisome Technology Exports," Fortune 97 (May 22, 1978): 106.

5. Brian W. Small, "How to Transfer Technology," Management Today (September 1977): 19-26.

6. John Diebold, "Doing Business in Developing Countries," Financial Executive (December 1973): 25.

7. David Ford and Chris Ryan, "The Marketing of Technology," European Journal of Marketing (United Kingdom) 11 (November 6, 1977): 369–82.

8. Ibid.

9. Eugene M. Lang, "Technology Transfer: A Quick Route to Foreign Markets," Management Review (September 1973): 62.

10. James R. Basche, Jr., and Michael C. Duerr, "International Transfer of Technology: A Worldwide Survey of Chief Executives," Conference Board Report (New York: Conference Board, 1975).

11. Harvey W. Wallender, "Managing and Evaluating Technology Transfer in Manufacturing Industries," UN Economic and Social Council, (Seminar on the Management of the Transfer of Technology within Industrial Cooperation, Geneva, July 14-17, 1975) Sc. Tech/Sem. 3/R.16. 13 June 1975.

12. R. G. Hawkins and B. Finn, "Regulation of Multinational Firms' Foreign Activities: Home Country Policies and Concerns," Journal of Contemporary Business 6 (Autumn 1977): 7-30.

13. W. W. Winpisinger, "The Case Against Exporting U.S. Technology," Research Management 21 (March 1978): 20.

14. T. Wilson, "To Maintain the Aerospace Industry's Economic Health," Vital Speeches of the Day 41 (September 1, 1975): 687.

15. M. G. Harvey and R. A. Kerin, "Multinational Corporations Versus Organized Labor: Divergent Views on Domestic Unemployment," California Management Review 18 (Spring 1976): 5-13.

16. Charles P. Lecht, "Tsunami," Computerworld 12 (June 12, 1978): 49.

17. Ibid.

18. Meyer, "Technology Exports," p. 106.

19. "French Exports See Future Competition," Business Week, no. 2484 (May 23, 1977), p. 34.

20. Lang, "Technology Transfer," p. 63.

21. "The Profitable Lure of Exporting Innovation," Business Week, no. 2586 (Industrial Edition) (May 21, 1979), pp. 40J, 40N.

22. "U.S. Must Improve Rules for Trade with Eastern Europe," Marketing News 10 (April 22, 1977): 11-12.

23. John Kiser, "Is the West's Technology 'Lead' an Illusion?," Government Executive 11 (March 1979): 39–40.

24. B. Richman, "Multinational Corporations and Communist Nations," Management International Review 16 (November 3, 1976).

25. W. A. Root, "Controls on West-to-East Technology Transfer," Defense Systems Management Review 2 (Winter 1979): 46.

26. W. S. Anderson, "Technology Transfer in Transition," Computers and People 27 (March 1978): 11-12.

27. Fred J. Bucy, "Technology Transfer: We're Still Selling Them the Rope," Government Executive 10 (September 1978): 38–39.

28. "Regulatory Maze Seen Problem for U.S. Firms," *Computerworld* 12 (January 16, 1978): 71.

29. Ibid.

30. "Export Controls Split Business," *Business Week*, no. 2584 (May 7, 1979), p. 36.

31. Richard Capstick, "The Perils of Manufacturing Abroad," *International Management* 33 (March 1978): 43.

32. Theodore H. Moran, "Multinational Corporations and the Changing Structure of Industries Supplying Industrial Commodities," *Journal of Contemporary Business*, Autumn 1977, pp. 121-32.

33. Cordell W. Hull, "Practical Considerations of Middle East Service Agreement," *Intertrade Journal* (June-July 1976): 24-26.

34. "Multinationals Find the Going Rougher," *Business Week*, July 14, 1975, pp. 64-66.

35. Bureau of International Commerce, "U.S. Companies Turning to Overseas Licensing of Industrial Know-How," *Commerce Today*, June 12, 1972, pp. 7-10.

36. "Latin America Opens the Door to Foreign Investment Again," *Business Week*, no. 2444 (August 9, 1976), pp. 35-36.

37. J. S. Fforde, *An International Trade in Managerial Skills* (Oxford: Basil Blackwell, 1957), p. 2.

38. Ibid., chap. 4.

39. Ibid., chap. 5.

40. Peter P. Gabriel, *The International Transfer of Corporate Skills: Management Contracts in Less Developed Countries* (Boston: Harvard University Press, 1967).

41. Ibid.

42. Richard Ellison, "An Alternative to Direct Investment Abroad," *International Management*, June 1976, pp. 25-27.

43. Ibid.

44. Clarence J. Mann, "The Technology of Mass Merchandising: Its Contribution to Economic Development," *Fund for Multinational Management Education* and the *International Chamber of Commerce* (Seminar on International Issues in Transfer of Technology, September 21-23, 1977, Paris, France), paper presentation.

45. H. J. Rath, "Management Techniques in International Technology Transfers," *Journal of Business Communication* 13 (Winter 1976): 25.

46. Kenneth A. Kovaly, "Market for Technology Keeps Getting Hotter," *Industry Week*, July 1, 1974, pp. 36-39.

47. Business International, *International Licensing: Opportunities and Challenges in Worldwide Technology Management* (New York: Business International Corporation, 1977).

48. Bureau of International Commerce, "Overseas Licensing," pp. 7-10.

49. U.S., Department of Commerce, Industry and Trade Administration, Office of East-West Trade Policy and Planning, Selected Trade and Economic Data of the Centrally Planned Economies, June 1979, p. 1.

50. Ibid., p. iii.

51. Ibid.

52. U.S., Department of Commerce, Industry and Trade Administration, Office of East-West Trade Policy and Planning, U.S. Trade Status with Communist Countries, November 15, 1979.

53. Irene Lange and James F. Elliot, "U.S. Role in East-West Trade: An Appraisal," Journal of International Business Studies 8 (Fall/Winter 1977): 5.

54. Ibid.

55. James R. Basche, Jr., "Evolving Corporate Policy and Organization for East-West Trade," Conference Board Report (New York: Conference Board, 1974), Report no. 635.

56. U.S., Department of Commerce, Selected Trade and Economic Data of the Centrally Planned Economies.

57. Charles S. Mayer, "Marketing in Eastern European Socialist Countries," University of Michigan Business Review, January 1976, pp. 16-21; David Clutterbuck, "Setting Up Shop in the East Bloc," International Management, February 1975, pp. 4-44; Jeffrey M. Hertzfeld, "Setting Up Shop in Moscow," Harvard Business Review, September-October 1974, pp. 137-42; John I. Huhs, "Developing Trade with the Soviet Union," Columbia Journal of World Business, Fall 1973, pp. 116-30; David Jones, "The 'Extra-Costs' in Europe's Biggest Synthetic Fiber Complex at Mogilev, U.S.S.R.," Worldwide R&I, May-June 1973, pp. 30-35; Robert E. Weigand, "Selling Soviet Buyers," Michigan State University Business Topics, Spring 1976, pp. 15-21; Leon Zurawicki, "The Cooperation of the Socialist State with the MNC's," Columbia Journal of World Business, Spring 1975, pp. 109-15.

58. Columbia Journal of World Business, December 1973, pp. 39-77, 83-89, 113-34.

59. "Soviet Trade: The Disappointments Continue," Business Week, December 6, 1976, p. 100; Herbert E. Meyer, "Why Business Has a Stake in Keeping Sovietology Alive," Fortune, September 1975, pp. 156-62; Herbert E. Meyer, "Why that Soviet Buying Spree Won't Last," Fortune, January 1975, pp. 94-138; Daniel Seligman, "Communist Ideology and Soviet Power," Fortune, January 1976, pp. 112-92.

60. Peter Gall and Lewis H. Young, "How Russia Would Spur Trade with the West—Interview with Nikolai S. Patolichev," Business Week, December 27, 1976, p. 34; Joseph N. Spiers, "More Technology Likely in Imports from USSR," Industry Week, November 3,

1975, p. 37; "Soviets Bring in Locals to Tell Approaches to Export Markets," *Industrial Marketing*, January 1976, p. 26; Joseph N. Spiers, "Technology Exchanges: A Route to Russian Markets?," *Industry Week*, April 14, 1975, pp. 7-8; "You Don't Sell Soviets—They Have a Shopping List," *Industrial Marketing*, July 1976, p. 6.

61. John B. Oliver and Elliot J. Weiss, "Is Selling Technology to the Soviets Dangerous?" *Harvard Business Review*, January-February 1975, p. 18; "Should Communists Buy US Technology?" *Industry Week*, December 1, 1975, p. 3.

62. Richard J. Janssen, "Debt Threat: Soviet Bloc Borrowings from the West Surge, Amid Mystery, Fears," *Wall Street Journal*, February 22, 1977, p. 1.

63. Business International, "Corporate Strategy, Planning, Organization and Personnel Practices for Eastern Europe," *Business International S.A.*, A Group Research Study, November 1977 (confidential copy of summarized results for sponsors only).

64. G. Peter Lauter and Paul M. Dickie, "Multinational Corporations in Eastern European Socialist Economies," *Journal of Marketing*, October 1975, pp. 40-46.

65. Business International, *Doing Business with Eastern Europe: A Guide for Western Corporations* (Geneva: Business International Corp., 1972).

66. Edward P. White, ed., *United States/Eastern Europe: Technology and Patents, Sale and/or Licensing* (Licensing Executives Society [USA] 1977).

67. Lawrance C. McQuade, *East-West Trade: Managing Encounter and Accommodation* (Boulder, Colo.: Westview Press, 1977).

68. G. Peter Lauter, "East-West Trade Organizational Structures and Problems of U.S. Multinational Corporations." Paper presented to the 3rd United States-Romanian Seminar on Trade and Economic Cooperation, Bucharest, Romania, July 17-22, 1978.

69. Ibid., p. 33.

70. Ibid.

71. Thomas Fahey, "Trading and Investing in Eastern Europe: A Businessman's Perspective," John K. Ryans, Jr., Anant R. Negandhi, and James C. Baker, pp. 132-42.

72. U.S., Department of Commerce, Industry and Trade Administration, *Working OnSite in Eastern Europe and the Soviet Union*, June 1977.

73. U.S., Department of Commerce, Domestic and International Business Administration, Bureau of East-West Trade, *Joint Venture Agreements in Romania: Background for Implementation*, prepared by Pompiliu Verzariu, Jr., and Jay A. Blurgess, June 1977.

74. U.S., Department of Commerce, Domestic and International Business Administration, Bureau of East-West Trade, Office of East-

West Policy and Planning, East-West Trade Financing: An Introductory Guide, prepared by Suzanne F. Porter, September 1976.

75. "Countertrade: Not for Everyone but Worth a Look," Commerce America, June 19, 1978, pp. 2-4; U.S., Department of Commerce, Industry and Trade Administration, East-West Countertrade Practices: An Introductory Guide for Business, prepared by Pompiliu Verzariu, Scott Bozek, and JeNelle Matheson, August 1978.

76. U.S., Department of Commerce, Bureau of East-West Trade, Countertrade Practices in Eastern Europe, April 1977.

77. "Guide to Trade and Cooperation with Eastern Europe," Berlin Chamber of Commerce and Industry (1 Berlin 12, Hardenbergstrasse 16-18, Germany, 1974).

78. U.S., Library of Congress, Congressional Research Service, Western Technology and Economic Performance in the Eastern Countries, prepared for the House International Relations Committee Hearings on Export Administration by John P. Hardt and George D. Holliday, June 1976. HC 56 Europe, East.

79. Robert W. Clawson and William F. Kolarik, Jr., "Trade Technology and Soviet R&D: U.S. Corporate Executives' View." Paper presented at the 1976 AAACS Annual Conference Panel "The State of Soviet Research and Development," October 8, 1976, Kent State University, Kent, Ohio, pp. 47-48.

80. John W. Kizer III, "Report on the Potential for Technology Transfer from the Soviet Union to the United States." Study prepared for the Department of State and the National Science Foundation, under Contract No. 1722-620217, July 1977.

81. Ibid.

82. Robert D. Schmidt, "Technology Transfer: More Rope or Which Rope?" Government Executive 10 (December 1978): 18-21.

83. U.S., Department of Commerce, Industry and Trade Administration, Office of East-West Policy and Planning, Bureau of East-West Trade, "Impact of Soviet Ministry Management Practices on the Assimilation of Imported Process Technology," (with examples from the motor vehicle sector), prepared by John P. Young, April 7, 1978.

84. "Technology Transfer—Is the U.S. Giving Away the Farm?" Government Executive VE 9 (July 1977): 25.

85. William C. Norris, "High Technology Trade with the Communists," Datamation 24 (January 1978): 99-103.

86. Peter Wiles, "World Technological Leadership," Lloyds Bank Review (United Kingdom), no. 127 (January 1978), pp. 15-27.

87. Jake Kirchner, "Study Sees No Cause to Bar Data Processing Exports to East," Computerworld 12 (October 16, 1978): 9.

88. Maurice J. Mountain, "The Continuing Complexities of Technology Transfer," Government Executive 2 (January 1979): 42-52.

89. Robert M. Bleiberg, "Dirty Business: The U.S. Should Ban Technology Transfers to the Soviets," Barron's 58 (July 31, 1978): 7.

90. Mountain, "Continuing Complexities."

91. Samuel P. Huntington, "Trade, Technology and Leverage: Economic Diplomacy," Foreign Policy, no. 32 (Fall 1978), p. 66.

92. Ibid.

93. Ibid., p. 67.

94. Eric W. Hayden, Technology Transfer to Eastern Europe: U.S. Corporate Experience (New York: Praeger, 1976), p. 23.

95. Ibid.

96. Paul Marer and Joseph C. Miller, "U.S. Participation in East-West Industrial Cooperation Agreements," Journal of International Business Studies 8 (Fall/Winter 1977): 17-29.

97. Ibid.

98. Business International, "Corporate Strategy, Planning, Organization and Personnel Practices for Eastern Europe," p. 1.

99. Ibid., p. 1 of summary.

2

METHODOLOGY

OBJECTIVES OF THE RESEARCH

An increasing number of corporations are seeking access to global markets. There is an increased need for skill in alternative forms of marketing. Knowledge that helps develop these skills is in demand. One objective of this research is to explore the technology transfer approach used when a business agreement is not accompanied by any equity involvement on the part of the transferring agent.

A second objective is to identify procedures or processes that may enable firms to accomplish such technology transfers more efficiently. Potential problem areas have been identified so that these may be avoided or revised. Analysis of selected cases should reveal those considerations that may improve the process of nonequity technology transfer in the future.

SIGNIFICANCE OF STUDY

A commitment to the nonequity technology transfer approach involves a shift for most companies from that of producing a product to that of training and supervising others to produce that product. This shift carries benefits, but also some disadvantages. It utilizes multinational corporation (MNC) technology in many cases, but it also deprives the MNC of certain employees, at least for the period of time during which the transfer training is taking place. In many cases, additional staff are needed so that the MNC's other production is not curtailed while the transfer is being carried out. Thus, a future commitment to additional transfers is often made so as to utilize this additional staff.

To choose not to become involved with technology transfer is not a simple decision, either. Potential markets may be avoided or overlooked by an MNC. This could be a deprivation and may not only affect the MNC but also potential host countries.

Definitions of successful transfers vary with the project. If some expectations are met, then at least an initial confidence in the operation and an acceptance of the technique are usually established. Exceeding expectations should not only provide support for the nonequity form of transfer but also build confidence for entering into similar future linkages. If the outcomes fall far short of the expectations, the cost could be high, depending upon the commitments that were made. The risks faced by MNCs in the nonequity transfers of technology and/or management are in the realm of corporate reputation, growth, and policy, as well as in the time and man-hours involved. These may not be as visible as a financial loss in an equity investment, but they may easily comprise an equivalent or an even greater detriment.

DEFINITIONS

A nonequity transfer arrangement is one that for the purposes of this study, does not require an equity involvement either initially or in the future. The time period is established contractually and is of a specific duration. The technology involved, in most cases, goes beyond the information or services usually provided in customary licensing agreements.

Technology transfer is an area subject to some confusion. For the purposes of this study, technology transfer will be defined as the transfer of hardware, such as factories, machines, products, and infrastructures (roads, laboratories, water distribution systems, storage facilities, and the like), and the transfer of software, including such nonmaterial ingredients as knowledge, know-how, experience, education, and organizational forms. Key know-how, such as the capacity to manage and apply techniques, is also included.[1]

It should be emphasized, however, that know-how may be even one step further removed, as pointed out by the American Economic Association, which has noted that technology "includes both specific knowledge about physical production embodied, for example, in machines, designs, formulas, etc. and also more general knowhows such as organizational procedures concerning coordination in production, procurement, marketing and research and development."[2] These definitions allow for many variations of nontraditional transfers. To avoid confusion, the types of involvements considered in this study are listed below. It is expected that most of the cases studied would encompass several of these involvements in the course of the project, as none of these are to be viewed as mutually exclusive.

1. Turn-key: the construction and equipping of a factory, plant, or place of operation only—although sometimes this term refers to the staffing and operation of a project. The accompanying services required to operate the plant once the initial switch is turned on or the plant put into operation will be considered separately.

2. Product technology: the product information that is necessary to create the product. However, it does not include the information necessary to construct or assemble the product.

3. Process technology: the information that details the assembly or construction of the product, often referred to as know-how. Another element that may be applicable here would be the adaptation of manufacturing methods to a revised version of the product.

4. Management (running the plant): this category comprises the initial start-up, including planning and proposal development, as well as operational aspects once the plant is functioning. It would include staffing but not the training of locals or in-country nationals. It would also include analysis of the operation, such as consideration of environmental support or other related issues.

5. Management (training): all training aspects, particularly in operations that are expected to become totally localized within a certain time period. The training may be entirely within the host country or may include training at other locations, including the MNC's headquarters.

6. Patent rights: patent rights are often included as part of the product technology, but it is possible to supply one without the other. In some instances the patent rights are filed long before the transfers occur.

7. Trademark rights: the right to the firm's logo as well as its technology, at least for a certain period of time, would be particularly important when payment for the services provided was to be supplied via the repurchase or buy-back arrangement. However, it would also be especially desirable in cases where the trademark would increase the demand for the product in that area. An accompanying factor to trademark rights would be inspection tours to ensure and maintain product quality.

METHODOLOGY

Description of Study

This study will look only at accepted agreements and the negotiation processes used to arrive at those agreements. Certain elements change after a contract is signed and accepted, and these will be detailed. Those factors that vary between the anticipated results and the final outcome will be identified with the intention of overcoming those variations in future transactions. Identification of the accompanying reasons for these discrepancies should help future transfers. In line with this, a comparison of the expectations and outcomes of selected technology transfers will be made by describing the actual outcomes of nonequity transfers of technology and/or management skills in their final forms at the project's conclusion, or at a certain stage if the project is still in process. These outcomes will be compared with the anticipated results projected at the start of the agreements. Transfers of technology and/or management skills in the Eastern European bloc, with its socialistic political systems, will be examined for two reasons. First, the political systems in most of these countries prohibit joint ventures involving an infusion of equity from the company based in the capitalist sphere; hence, nonequity agreements are more common in this region. Second, prior studies, particularly the research conducted by Eric Hayden, have established a groundwork by describing some of the expectations that existed at one point in time. Building upon this base offers more opportunity for an in-depth study than would be the case if other transfer examples were selected.

Nonequity transfers will be examined by looking at aspects prevalent in the selected agreements. Information has been collected—much of it through personal interviews—from MNCs with current or prior experience with the nonequity transfer mode. The firms cooperating on this research have had nonequity transactions in one or more Eastern European countries. They are listed in Appendix A in detail, but include the following U.S. companies and their foreign partners: Clark Equipment Company—Bumar Foreign Trade Enterprise (Poland) and Universal Tractor Foreign Trade Company (Romania); International Harvester Company—Bumar Foreign Trade Enterprise (Poland) and Hungarian Railway Carriage and Machine Works (Hungary); Honeywell, Incorporated—Mera/Metronex (Poland); Westinghouse Electric Corporation (Semiconductor Division)—Unitra Electronic Industry Union (Poland); General Tire International Company—Romania State Enterprise of Foreign Trade (Romania); The Bendix Corporation—Varazeinski Tekstini Komdinat (Yugoslavia).

Initial interviews with several corporations indicated that, in some cases, several corporate officers would need to be contacted with regard to a project. These decision makers were usually at the vice-presidency level, but day-to-day handling of most transfer projects was done on the managerial or program coordinator level. Supporting information obtained from these employees, particularly managers involved with the projects, contributed to the case studies described herein.

Information obtained through personal interviews has been structured along the format of the actual outcomes (or the situation in progress) and compared with the expectations that were said to exist prior to the start of the transfer situation. The questionnaire that was followed during the personal interviews, including the "actual versus anticipated" results format, can be found in Appendix B.

Selection of Cases

The Eastern European countries selected for this study, and the further choice of corporate projects within those countries, were chosen with several reasons in mind. As stated earlier, this work was designed to complement the earlier research of Hayden, as far as possible. This was done primarily to take advantage of the data on agreements that had been already collected and analyzed. Furthermore, since much of the data represented opinions and expectations at a point in time near the initiation of many of the agreements, it provided a secondary indication of corporate anticipations. The companies selected for this study were drawn from among those in Hayden's sample that were willing to participate in this research and share their observations regarding the outcomes of their projects. "Outcome" refers here to conditions existing either at the conclusion of the project or, for an agreement in operation, at a point in time substantially after start-up.

The arrangement between Yugoslavia and The Bendix Company represents a significant point of departure from the above rationale and was sought out for two reasons. First, Yugoslavia, like Romania, does permit some equity ventures to be established. Thus, this agreement provided a similar background in the direct investment sense to the study of Romania. Second, since Bendix had been in the process of establishing a major nonequity venture with the Soviet Union, the company's actions indicated an acceptance of the nonequity format and a desire to establish contacts within the Soviet bloc. Since the contract with the Soviet Union had been "postponed indefinitely" just prior to this study, the company's assessment of the nonequity approach promised to be interesting.

Research Areas

This study encompasses two basic research areas: (1) descriptive factors and (2) analytical factors. The descriptive factors highlight aspects of the respective transfer projects and, as the term implies, are descriptive in nature. As such, they are essential to understanding the total scope of such nonequity agreements. The analytical factors are elements that emerged in the course of the study and were seen to have a major impact upon the attainment of corporate expectations.

The following factors evolved as descriptive of nonequity transfer agreements:

Direct transfer
Indirect transfer
Environmental factors
Political factors
Financial factors
Leverage
Contract format
Expected events that did not occur
Unexpected occurrences
Anticipated benefits to the corporation

A discussion of each of these ten descriptive factors follows and is based on information obtained from a review of literature currently available and some preliminary interviews with the corporations.

Direct Transfer of Technology

Based on our definition of technology transfer, direct transfer encompasses all information involved in the transfer—whether tangible enough to be found on blueprints or as intangible as human experience and know-how.

Hayden noted some of the adverse factors that MNCs expected prior to such undertakings. These included awkward extant designs, poor production techniques, low quality control standards, and very low worker morale and initiative.[3] However, he pointed out that expectations of such conditions did not always include accurate preconceptions as to the extent of the conditions.

Several of Hayden's points allude to worker skills and training tasks, which are part of the direct technology transfer. Language proficiency—especially on the part of the workers to enable them to receive certain training—is also a factor. Without adequate communication of training skills, the mastery of processes and techniques would be affected.

Harvey Wallender, in his paper "Managing and Evaluating Technology Transfer in Manufacturing Industries," made an observation that explained much of the disparity between a company's expectations and outcomes—technology was often transferred as a bulk process when it should have been done in stages, over a period of time, with different approaches being used at each stage.[4]

Fforde detailed some problems in the marketing/administration and production areas. He found expectations in the marketing/administrative area often in error as to the market demand or potential, or as to the competency of the local staff (often hired by the host party). This latter aspect tended to heighten disparity in the second area, production, when it was below or behind expectations. This was often due to training difficulties, sociocultural shifts, or technological aspects on the job.

Fforde did not believe that the lack of competent staffing was solely the fault of the host party, however; he noted that often many capable employees did not want overseas assignments. As these individuals often possessed enough clout to turn down overseas positions, less competent individuals were permitted to assume the posts.[5]

A brief telephone interview with Honeywell, International, confirmed another aspect of expectations first noted by Fforde—the lack of skill or appropriate background of local workers. As a precautionary measure, Honeywell had attempted to select a level of technology appropriate to the ability of the host party. Inappropriate selection, it was felt, might result in the selection of a technology that would be unable to be transferred within the project's time frame with the requisite level of quality-controlled production.

As can be seen, a comprehensive description of the immediate aspects of a direct transfer of technology would include detailing the products and processes of the technology, the skills and training tasks required of both the trainers and trainees, an assessment of a common language area, consideration of delays directly related to the production areas, and other technologically related factors.

Indirect Transfer of Technology

The indirect transfer of technology is connected to direct transfers but is one step removed from the specific technology of the latter in that it deals with the supporting needs that are required in the transfer. Certain transfers described by Hayden, Fforde, and others allude to the equipment purchased from the transferring company or the plant construction and layout designed by the company selling the technology but implemented by the purchasing party.

Another aspect of indirect transfers involves the acquisition of raw materials, since part of the justification for onsite manufacturing of the acquired products implies the use of local raw materials. In such instances, any services rendered by the transferring company along the lines of raw material procurement or quality testing would impact upon the technology transfer.

Therefore, the indirect transfer of technology includes such issues as construction, equipping, and layout of any and all plants or factories used in connection with the purchased technology, raw materials acquisition and approvals, and other supportive capabilities.

Environmental Considerations

Environmental factors in a technology transfer deal with the surroundings in which the transfer takes place. For the most part, this relates to the host country or territory, but it might also include the home base of the transferring agent. In the strictest connotation of the term, the environment refers to a geographic area with its accompanying climatic, topographic, and other factors. In a broader sense, however, the environment relates to the sociocultural milieu within which the technology transfer occurs. Hayden's description of low morale or worker apathy might easily fall into this realm.

Since the transfers considered here involve Eastern Europe, security issues are also a factor to be considered in their description. Restrictions placed upon the seller of technology with regard to its application, or even implementation, could affect the efficiency as well as the overall outcome of such a transfer.

The larger technology transfers, especially those requiring construction or lengthy stays within the Eastern European country, must also have an interface with the transportation systems of the country. Also, those companies agreeing to have relatively lengthy incountry stays for their executives and representatives will also be concerned about the accommodations available to their employees. In a preliminary interview, Clark Equipment Company noted that plans for MNC staff accommodations abroad should include appropriate housing conditions for family members—but that different definitions of *appropriate* might occur.

The category of environmental considerations, therefore, includes geographic factors as well as sociocultural interfaces, security issues, logistics, and accommodation factors, which comprise the surroundings within which the transfer is to occur.

Political Arena

The security issue was included under the heading of environmental considerations; apart from this, however, a separate category of political factors considers the major political interfaces that might occur, particularly when the technology transfer is taking place between entities based on or representing differing political ideologies.

Certain technologies in the United States are controlled by the federal government and are not permitted to be sold to all countries, particularly those connected to the socialist bloc. Thus, one issue in this category would be home governmental approval and permission to effect such technology transfers to Eastern Europe.

On the other hand, governments in Eastern Europe may have similar if not equal objections to trading with companies based in a capitalist political sphere. The restriction against equity-involved joint ventures in many of the Eastern European countries is one indication of such restrictive policies.

In general, labor union objections to the idea of job export might also be viewed as a particular type of political issue, since some of those union concerns could be substantiated by legislation. Therefore, this aspect of technology transfer should be examined, if only for its impact on current and future transfers.

The political arena category covers both home and host parties and their respective governmental issues, and home labor unions and their possible objections to the export of jobs or to other aspects of trade with socialist bloc nations.

Financial Flows

Regardless of the rationales taken by corporations in their individual justifications for accepting technology transfer arrangements, certain financial flows are assumed to be integral to the success of the projects. Therefore, this category of financial factors examines the approaches taken to payment in each transfer.

The most obvious consideration here would be the payment arrangements laid out by the corporations, with the accompanying payment schedules. Supplementing these payments would be additional purchases, made by the host company or country, either of total units (potential substitutes for the technology purchased until local production was in operation), or of subassemblies (to aid production and help with training), or otherwise, such as raw materials.

Incidental expenses encountered by both parties would be another issue, and this could refer to additional expenses encountered in equipment purchases or in construction aspects (especially if these areas were part of the transferring agent's responsibility) or expenses connected to housing or transportation or other areas.

Since, in each case, the MNC is selling its technology for a certain price, payment for this technology is a crucial aspect of each of these transfer arrangements. The category of financial flows looks at payment schedule commitments, supplemental purchases—especially in the area of products prior to on-stream local production—and incidental expenses.

Leverage and Control

The category of leverage and control grows out of the possibility that a sale of technology without an accompanying equity involvement might place the seller in an uncomfortable position if the information had been provided but the payment had not been made. Whether or not an equity basis does indeed provide adequate leverage (particularly to the minority partner) might be debated, but the lack of any such mutual involvement requires the seller of the technology to gain control of integral portions of the operation in other ways. Negotiating leverage should be relatively high for the seller at first, especially if the technology has been sought out by the host company or country; however, perpetuating this leverage-type position appears to require additional effort.

Ellison felt that acceptance of the nonequity transfer arrangement carried over into the control issue—that is, MNCs that supported the nontraditional approach by choice (as contrasted to those involved in expropriations) also cooperated in the area of shared control more easily.[6]

Several MNCs in Ellison's study were involved in an equity participation arrangement. He found this aspect to be detrimental to decision making, an opinion contrary to most traditional thought in this regard. Ellison noted that fees were often unsatisfactory with regard to expected profits and there appeared to be a lack of guidelines for the establishment of fees in general.[7]

Attitudes toward "control" varied from company to company. Some firms felt that leverage or control in the shape of a "missing technological link" for the product was vital—that is, holding back a portion of the information until other commitments had been completed. Others noted that a firm's marketing expertise for the completed products was enough to act as a control mechanism. Still others noted that just the desire for a continued association, including updated technology and additional contracts of the nontraditional transfer form, provided enough leverage to make the relationship successful. In any case, examination of this category leads directly into the next: contract format, which is designed to enhance or at least preserve a firm's leverage.

Contract Format

Another area that is a potential source of difficulty is the type of contract drawn up for the agreement. Ellison prefers the shorter, simpler document because he feels the lengthier one inevitably excludes some item.

He also urged that the training of nationals be as rapid as possible. He pointed out that countries often saw this training as the most important benefit of the transfer, while MNCs viewed production of the product as the chief aim.[8]

Cordell Hull, vice-president of Bechtel, International, offered suggestions regarding Middle Eastern agreements, but his suggestions would apply to most areas of the world. He pointed out the major areas that give rise to difficulties in meeting expectations.

1. The time estimated to accomplish transactions is too short (should be three to four times longer).
2. There is little to no recourse to laws outside the limits of the contract itself.
3. Multiple language usage, when necessary in creating an agreement, should be checked for accurate translation.
4. Cost-plus-fee contracts are more difficult compared with fixed-price contracts.
5. Multiple currency components should be set up with an established rate of exchange so that monetary fluctuations would be partially reduced.
6. Insurance often takes longer to obtain and costs more than expected.
7. "Unexpectancies" suggested placing a force majeure clause in in the contract.
8. Arbitration, suspension, and termination clauses in the contract should be thoroughly reviewable.
9. Realize that guarantees may not be secure or forthcoming.
10. Include a detailed description of all performances expected by both parties, since this is an area that could be easily misinterpreted.[9]

Hull's listing covers many areas wherein outcomes did not keep pace with expectations, but he also provides suggestions that sharpen one's awareness of those discrepancies. However, it does appear difficult to satisfactorily include all the details recommended by Hull in the short, simple contract form suggested by Ellison in his research.

Gabriel, who found general expectations matching outcomes in a all four of his situations (actually, outcomes in two examples exceeded expectations), made several observations that ought to be studied before one enters into nonequity transfer agreements. They include the following:

(1) risks involved in starting such technology transfers includes the number of employees since the additional staffing requirements provide at least a partial obligation toward similar commitments in the future;
(2) agreements with governmental concerns carry problems which private firm contracts do not usually have, and thus, such agreements should be carefully scrutinized on this basis;

(3) the most powerful position of the MNC is at the start of the undertaking, or during the negotiations, hence, any desired changes are best enacted at these times, not afterwards.[10]

Corporate experiences have amplified the above observations. Clark reported instances in bargaining when signed contracts were not accepted as an accomplished fact. It stressed the point that some nations might be reluctant to follow the contract exactly.

The Bendix Corporation cited an initial hesitancy regarding concluding negotiations since the "assumed concluded" agreement to establish a spark plug plant in the USSR had just been postponed indefinitely. The firm also noted that countries often try to establish contracts in such a way that the MNC is responsible for all unexpected occurrences. Officials noted that one should avoid contracts worded in such a way that states that if the MNC's end of the agreement has not been fully completed within the specified time period, the financial payments owed by the host country will never be completely forthcoming.

International Harvester noted differences in the honor accorded a contract in socialist countries as compared with some other countries. Several individuals involved with International Harvester negotiations expressed the view that "one may negotiate long and hard to obtain a contract in the socialist state, but once it is signed, it is honored; in other countries, this is not always so"—a benefit of dealing with the socialist bloc, in their opinion.

Expectations That Do Not Materialize

In nonequity technology transfers, most of the essential points of the agreements appear to be fairly clear-cut: the technology involved, the amount and timing of the payments, the time frame of the agreement. However, certain expectations might exist that would not be fulfilled. Hayden lists particular problems that may exceed the specific expectations of the parties involved, although these were not seen as insurmountable. His list includes the following:

U.S. governmental approval delays
Equipment delivery problems
Worker apathy
Lack of worker technical skill
Consumption of more corporate time and energy than expected[11]

Fforde found three general areas where actual outcomes often fell below initial expectations. The first area encompasses expectations that were based on the initial contact, when misconceptions were inadvertently established. Such misconceptions include the idea that the host party knows more about the operation, particularly regarding manufacturing expertise, than is actually the case; that the association between the MNC and the host is stronger than it actually is, simply because a highly knowledgeable MNC and a well-known local party are involved; and that the MNC can overcome cross-cultural difficulties because the MNC has accepted the assignment when, in reality, these difficulties are often beyond the power or awareness of the MNC.[12] Obviously, some expectations, large or small, will presumably never be totally fulfilled on every project. However, this category's purpose is to examine the expectations to determine, if possible, the reasons behind their nonfulfillment.

Unexpected Occurrences

A category covering unexpected occurrences would require the continual collection of information. However, it has been included here to aid in documenting aspects of contract agreements that are sometimes not anticipated but that may have had a major impact upon the project. One allusion to unexpected events was made by Ellison in the form of a recommendation rather than an itemization of unexpected possibilities. He recommended some restructuring within the MNC to allow for more flexibility in dealing with unexpected occurrences, such as manpower or staffing problems.[13] Preliminary interviews with Clark Equipment Company lent some insight into this area.

Executives at Clark Equipment stated that unexpected situations had arisen regarding the number of staff members required and/or the period of time for which their services were needed. On the other hand, it was pointed out that mandatory adherence to a schedule could jeopardize the project, while extended staffing could be a means for successful completion of the project. Thus, certain requirements should remain flexible.

Clark also pointed out another aspect concerning a problem that had been planned for even though it hopefully would not arise. A host country improved upon the basic technology of the MNC. When the MNC requested royalties, however, the MNC's right was questioned because the patent involved was not the exact one supplied by the MNC. Interpretation of the arbitration clauses as well as other portions in the contract were expected to be required before this issue was successfully settled.

One positive discovery that had exceeded all expectations of Clark Equipment was that reciprocal development of new products had been gained through nonequity technology transfers.

Thus, unexpected aspects might involve a variety of areas, including rapport between the parties involved, internal and external changes, product sourcing, quality controls, or technological boundaries and developments, among others.

Benefit to the Corporation

The financial flows category provides a means for examining monetary exchanges as laid out by the agreements, but it does not fully consider the reasons behind a corporation's decision to enter into a technology transfer. Profit may be the chief reason, but several other aspects appear to underlie the decision and may, in some instances, be the primary rationale.

Bendix Corporation stated that specific transfers were hoped to be profitable based on fees, as stated in the contract. Their expectations would still be met, however, even if the endeavor proved profitable only as to the sale of products or parts.

International Harvester (IH) representatives also stressed the point regarding payments and profits. When IH agreements permit full disclosure of the needed technology, a major source of income in the arrangements is often the inclusion of purchases of parts from IH on an ongoing basis during but also after the technology transfer has been completed.

Regarding management contracts in particular, Ellison noted that expectations varied with the initial reason underlying a firm's entrance into a nontraditional transfer agreement. He pointed out that "aggressive" firms fared more successfully and profitably than those "after-the-fact" firms that undertook such a transfer as a means of recouping some of their losses due to expropriation.[14]

The category of corporation benefits, therefore, looks at the major reasons behind a corporation's decision to enter into a specific technology transfer, ranking those reasons if feasible. The intention, then, is to compare the rationale behind corporate expectations with the benefits actually gained by the corporation from the transfer.

These ten descriptive categories provide a basis for comparing expectations and outcomes; they are summarized in Figure 2.1. An analysis of the resulting differences between expectations and outcomes will provide a list of analytical factors that, in turn, should provide some guidance for future technology transfers.

INTERVIEW GUIDE DEVELOPMENT

From the ten categories describing elements involved in nontraditional transfers, the question areas were selected. The layout of questions on the questionnaire was designed to elicit comparative

FIGURE 2.1

Descriptive Factors of Technology Transfer

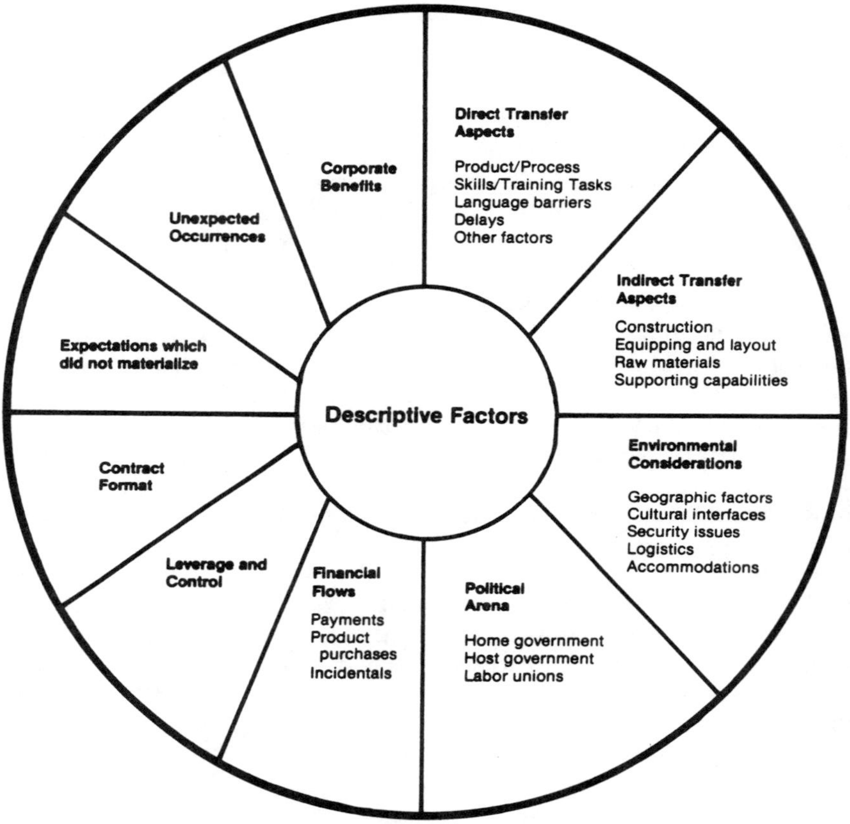

responses. The questionnaire was essentially an interview guide, as the information was obtained through personal interviews with corporate officials.

Background and Overview of Company, Country, and Product

A general section in the questionnaire was provided to obtain background information from the corporations. This was positioned at the beginning of the questionnaire (and, hence, the interview) for two reasons. First, it provided a perspective from which to compare the information that followed relative to the specific agreement that would be discussed. Second, such questions are often nonthreatening and carry little to no risk in terms of confidential information. Thus, it provided an opportunity to become acquainted on a low-risk yet informative basis. Such information was judged relevant to the study because it was assumed that perspectives on various projects might vary depending on past incidents or the lack of such experiences.

Title, Date, Descriptions of Specific Agreement(s) under Consideration

Questions concerning title, date, and agreement descriptors began the narrowing down process in concentrating on particular agreement(s). A checklist of variables such as turn-key, product-process transfer, management, patent, trademark rights, and buy-back provisions was designed to speed the information-gathering process yet not overlook aspects of the contract. Each company's arrangement with a single country was considered separately, unless comparisons were offered by those involved; in some instances, one general agreement was broken down into separate contracts. Hayden had distinguished such areas as technical assistance, trademarks, and buy-back stipulations; in situations where these areas were covered by separate contracts, the discussions still considered these aspects to be part of the overall arrangement between the two parties involved. Requesting a listing of corporate responsibilities as well as the responsibilities of the other party helped to clarify the extent of the technology being transferred—especially in situations where permission to study the actual corporate contract was denied. Since Eastern European industrial arrangements must be concluded between state trading enterprises (yet are often negotiated on the level of the ministries), the identification of each party or level involved in the transfer helped to overcome confusion as to the identification of the parties.

Negotiations

The impetus behind the desire for the agreement was seen to be a highly relevant factor, as pointed out both by prior studies as well as the preliminary corporate interviews. In addition, the actual negotiations were expected to be important, that is, as to the length of time involved, the initiator, and the type of format adopted for the official contract. Thus, questions regarding the negotiations were left fairly unstructured so that many individual points could be provided.

Time Frame

It was pointed out that delays were a major factor in the eventual success or failure of an agreement; hence, the time frame as originally projected would be compared to the actual timing. It was also important to distinguish the timing between different phases of the projects.

Aspects of Involvement

"Aspects of involvement" was used as an umbrella category covering financial flows, staffing and personnel arrangements, and the protection accorded to project participants. Financial flows were considered vital to each arrangement, since the profit-centered MNCs had undertaken the technology transfer under a sale basis. However, several aspects existed within the financial area. First, were the payments themselves, as contractually arranged. Second, these payments would not reflect the profitability of the venture without an accompanying statement on the expected or unexpected expenses incurred during the course of the project. Third, profits based on products or items expected to be sold during the course of the agreement would be another factor. Fourth, additional sales or financial flows not actually expected in connection with the project would also impact upon the overall evaluation of the financial picture of the technology transfer.

The personnel aspect was considered from two bases: employees of the U.S. company involved in the transfer and the trainees (those employees selected to receive any particular training necessitated by the transfer) from the host company or nation whose education, in most instances, was expected to be an integral part of the transfer. The personnel issues expected to most affect the domestic company included both the availability of key personnel and factors causing difficulty or hardship in connection with the transfer. Issues considered particularly crucial to the host involved training difficulties and quality controls. The underlying belief was that staffing impacted on financial flows because delays or an unsuccessful transfer would negatively affect the profitability of the venture.

Protection was an issue based in the control-leverage attitudes of direct investment agreements. A study of the leverage available to the MNC once the agreement was signed (which, at least according to one author, was then a period of greater risk and less control) was considered desirable for the construction of future recommendations as well as current analyses. Protection also involved issues affecting patents and trademarks in a security realm, as well as legal protection that might be available in court systems or arbitration proceedings.

Problems

The problems section on the questionnaire was included as "insurance" that items relevant to the transfer that might have been overlooked in prior categories would be noted. It was not intended to emphasize problems over opportunities or other aspects of the transfer. It did appear to be true, however, that areas that varied appreciably from expectations were often described as problems.

Assessment

Assessment questions looked at aspects affecting the ease or difficulty of carrying out the project. They provided an overview that was broad enough to touch on other aspects of the agreement that might not have been covered previously. The checklist approach to the opportunities involved in the technology sale was designed to emphasize points that might have been unspoken, especially if one or two major reasons behind the desire for the transfer were so dominant that some lesser reasons were not acknowledged adequately if at all.

Variations

The technique was to obtain an accurate description of each agreement, including any phases, and then to explore areas where variations had occurred between initial expectations and the final outcome (or the stage at which the transfer was currently involved). Distinguishing straight descriptions and the variations that occurred (or were perceived to have occurred) from initial expectations was seen as one way of isolating two separate points of analysis. Thus, the questionnaire sought to bring to light changes or variations as appropriate in each category.

LIMITATIONS OF METHODOLOGY

The approach utilized in this study should provide some pragmatic and useful information that will be of benefit to corporations considering technology transfers, particularly to countries within the socialist bloc. However, this methodology carries some limitations.

FIGURE 2.2

Descriptive Factors Leading to Analytical Factors in Technology Transfer

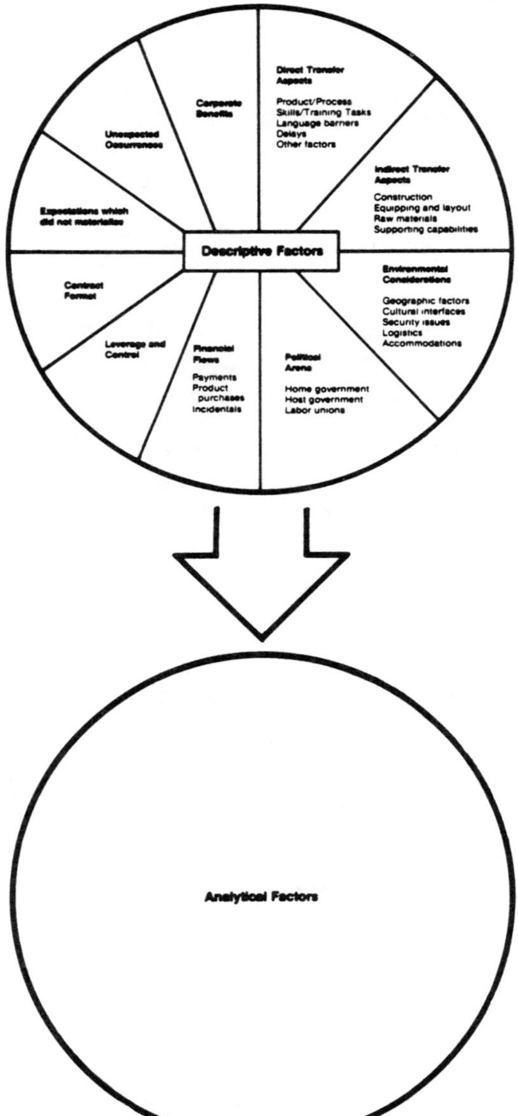

The extremely small number of corporations used as the basis for this information does not permit generalizations to be made to the experiences of more corporations. Personal interviews provide more information, on the whole, than short, specific questionnaires, but the opinions of the individuals involved represent only that person's viewpoint. A discussion of expectations, from hindsight in some instances, is often colored by events that have occurred since they were formed. Furthermore, corporate personages notwithstanding, expectations vary—sometimes radically—with different individuals, even though they are involved with the same corporation. Thus, information is limited by the number and bias inherent in the individuals providing the material, over and above all statistical information regarding each project. On the other hand, however, analyses of these technology transfers will also be made by the same individuals. These decision makers will determine, in part, future corporate transfers. Information that will improve corporate performance in this regard will benefit these individuals, the corporations, and the receiving host companies and countries.

Thus, an overview of the technology transfer as it will be seen in this study involves consideration of the initial criticisms of such transfers, but also the justifications. It examines technology transfers on a descriptive basis, but seeks the analytical factors that underlie those arrangements. Figure 2.2 illustrates the relationship between these elements.

NOTES

1. U.S., Chamber of Commerce, International Transfer of Technology Sources of Conflict, p. 5.
2. E. Hewett, "The Economics of East European Technology Imports from the West," American Economic Association, May 1975, p. 377.
3. Eric W. Hayden, Technology Transfer to Eastern Europe: U.S. Corporate Experience (New York: Praeger, 1976), pp. 54-56.
4. Harvey Wallender, "Managing and Evaluating Technology Transfer in Manufacturing Industries," UN Economic and Social Council (Sc. Tech./Sem. 3/R.16), June 13, 1975.
5. J. S. Fforde, An International Trade in Managerial Skills (Oxford: Basil Blackwell, 1957), chap. 4.
6. Richard Ellison, "An Alternative to Direct Investment Abroad," International Management, June 1976, pp. 25-27.
7. Ibid.
8. Ibid.

9. Cordell W. Hull, "Practical Considerations of Middle East Service Agreements," *Intertrade Journal*, June-July 1976, p. 25.

10. Peter P. Gabriel, *The International Transfer of Corporate Skills: Management Contracts in Less Developed Countries* (Boston: Harvard University Press, 1967).

11. Hayden, *Technology Transfer to Eastern Europe*, p. 55.

12. Fforde, *Managerial Skills*, chap. 4.

13. Ellison, "Direct Investment Abroad," pp. 25-27.

14. Ibid.

3

DESCRIPTION AND ANALYSIS OF CORPORATE EXPECTATIONS IN NONEQUITY TECHNOLOGY TRANSFER TO EASTERN EUROPE

INTRODUCTION

"Expectations" might well be the subtitle of this section, as the intention of Chapter 3 is to delineate the expectations of six corporations that established nonequity arrangements with countries in Eastern Europe. Two of the companies established similar agreements with two different countries, so eight different agreements are described in this chapter. Each agreement has been treated separately, but all have been formulated along the same structure.

Outline

The outline followed in describing corporate expectations is as follows:

COUNTRY

<u>Company</u>

Description
- Background of domestic company and host
- Transfer of technology
- Responsibilities

Corporate perception of benefits to be gained from agreement

Negotiations
- Parties involved
- Initiating contact
- Length of negotiations
- Issues and obstacles

Time Frame
- Overview
- Special considerations

Financial Flows
- Payments
- Purchases from the domestic company
- Purchases by the domestic company

Personnel
- Training tasks
- Other personnel involvement

Protection
- Leverage
- Inspection privileges
- Patent/trademark defense
- Legal provisions

Project assessment
- Factors expected to make the project easier than usual
- Factors expected to make the project more difficult than usual

Expectations for the future

Country-Company Arrangement

Four countries are represented in these descriptions. Two of the countries were involved with several of these projects simultaneously, while the other two countries had only one project considered in this study. Poland, by far, had the greatest concentration of corporate involvement as represented in this analysis. The companies and their respective products involved in technology transfers to Poland include the following: Clark Equipment Company—axles; International Harvester Company—complete construction vehicles; Honeywell, Incorporated—electronic pressure transmitters; and Westinghouse—semiconductors. In addition, Romania purchased technology for radial tire manufacture from General Tire Company and transmission technology from Clark Equipment. Hungary had a series of arrangements with International Harvester for the transfer of agricultural equipment technology, and Bendix Corporation was in the process of transferring the information required to manufacture material for disc brake pads to Yugoslavia.

This chapter is involved with expressing the expectations of each company as those expectations existed at the start of each agreement. Effort has been made to include the relevant descriptive details of each arrangement so that a more accurate comparison will be possible in the analysis with the subsequent outcomes.

POLAND

Clark Equipment Company

Description

The Clark Equipment Company, based in Buchanan, Michigan, signed an agreement in 1972 with the Bumar Foreign Trade Enterprise (FTE) in Poland to transfer Clark's technology relating to the manufacture of cast housings and axles. Through the Bumar FTE, Clark was associated with the Bumar Union of Construction Machinery Industry, which includes all Polish enterprises involved in earthmoving, construction, or road building undertakings in Poland.

The arrangement between Clark and Bumar was divided into three sections. Under the technical assistance portion, Clark agreed to transfer to Bumar all information relating to the manufacture of heavy-duty, cast-housing planetary reduction axles (drive axles, drive steer axles, and tandem axles).[1] The second portion consisted of trademark rights, which Clark agreed to transfer to Bumar for each of those axles. The third portion involved a marketing agreement in which Clark agreed to repurchase some of the axles from Bumar.

Background

Domestic Company. Clark's product line includes construction equipment, cranes, forklift trucks, axles, and transmissions or drive trains. More than 90 percent of Clark's present business is in axles, transmissions and related components, material-handling equipment, and construction machinery. Axles and transmissions accounted for \$299.9 million, material handling equipment for \$766.8 million, and construction machinery for \$665 million, for a total of \$1,731.7 million in worldwide sales in 1979.

Clark Equipment stated that it adapts its agreements to fit both the country and the tasks at hand. Thus, the corporation might have several rather unconventional types of arrangements functioning concurrently. This approach differs from many firms that tend to hold fast to one type of agreement, often that of 100 percent ownership or nothing. Clark has had numerous majority interest ventures but is also open to minority ones. For example, Clark has one minority venture with Mexico wherein Mexico controls 74 percent and Clark 26 percent, with loans backing this joint venture to an amount of nearly $14 million as of 1979. Clark has also been exploring a type of agreement known as product-in-hand, which is to be a combination of a turn-key venture combined with a management contract, in addition to the usual technology product and process transfer. Clark officials summarize such a project as a situation wherein the company staffs and runs the plant as well as builds it. The contract requires the corporation to guarantee a certain level of output, and, in some cases, the corporation's payment is taken at least partly in the form of that plant's production. Perhaps the most obvious difficulty in such an undertaking is the enormous size and responsibility of such an endeavor—not to mention the number of things that could go wrong prior to the time production is on-stream and meeting projected quotas. Clark noted that some countries seem to be making this product-in-hand type of arrangement a national policy, but also noted that a country must appear to offer great opportunities, financial or otherwise, often market entry or raw material acquisition, to have the necessary leverage to induce a corporation to accept such a large responsibility.

The history of Clark Equipment's overseas involvements began after World War II when Clark, a domestic company with not overly large sales, found that parts, repair, and maintenance were needed for the forklift trucks sold during the war that were left all over the world. So Clark began franchising dealers in 1946 and soon had $5 million to $6 million in exports.

In the 1950s, Clark formed their international division and began licensing in Australia, Belgium, Germany, France, and England, forming joint ventures everywhere except in Belgium, where the company was usually the minority partner on a basis ranging from 30 to 45 percent ownership.

In the 1960s Clark began licensing in the larger developing countries such as Argentina and Brazil. The corporation went on buying sprees in England, France, Germany, and Australia, buying control of operations. New joint ventures were formed with substantial equity in Mexico, Belgium, Japan, and in South Africa on a minority basis of 20 percent.

In the 1970s the purchases of the 1960s were rationalized and refined. Some operations were closed down, such as the one in England, and the corporation went back to the development philosophy, licenses, and joint ventures of its 1950s position. But note that in the 1960s, Clark was able to take control because the company had prepared the way in the 1950s; and, in the 1970s it could do the same thing in the developing markets. This was the idea behind licenses in such countries as Korea, Taiwan, Algeria, and India. Potential agreements are still pending with Indonesia, Egypt, and Venezuela. Algeria was a possibility at one time but that has never been finalized. Licenses without investment were also established in New Zealand. Selection of countries was based on environmental risk ratings with preference toward the lower risk areas.

Clark feels one should not stop benefiting from what is already possessed. The company earned $870 million outside the United States, with direct sales of $599 million and licenses of $271 million as of 1979.

In 1945 the international aspect of Clark was limited to a network of domestic distributors that exported replacement parts. However, increasing worldwide competition eventually encouraged Clark to branch out to a system of licenses in which foreign companies assembled or manufactured Clark products, paying Clark for its services. To coordinate its international activities, Clark has an export marketing subsidiary, Clark International Marketing S.A., and a licensing subsidiary, Clark Equipment A.G., which handle all licensing concerns outside the United States. Clark is unique in that it adopted a minority position or even a nonequity basis for many of its agreements internationally far earlier than most companies.

Clark Equipment stated that it has a high commitment to the international aspects of its business, with a corporate goal at one point in time to have at least 50 percent of its business conducted overseas by 1980. However, because the corporation is aware of its limitations, financial and otherwise, it is probably even more cautious than other larger companies such as Caterpillar Tractor or General Motors.

Clark's approach in the transfer of technology is to use know-how packages to form a wedge into the market with their partner. Clark sees local production of their products as a necessity in the distant future in terms of remaining cost-competitive with other firms.

International contracts of a licensing nature are handled by a separate division within the Clark Equipment Company. This international licensing division, which operates as a separate profit center, is the arm that transfers technology for Clark. The corporation feels that the culminative international experiences of this unit provide continuity in their overseas arrangements.

Clark's approach to the majority of its international license arrangements is that it provides all know-how as well as documentation to ensure that regardless of where in the world the product is produced, it will be manufactured with the same techniques and to the same standards that Clark demands of its domestic manufacturing operations. Supporting this tenet, Clark always supplies its trademark with such technical assistance agreements.[2]

Managerial skills and the transfer of technology as separate areas are blurred. Officials at Clark pointed out that they were know-how people, but that the application and implementation of that know-how also required managerial skills. Clark provides managerial assistance for the licensee because the company does as much as is feasible to make the licensing situation a success.

Clark feels that licenses are flexible, "living" documents, tailored to the needs of the particular situation. Clark has noticed that arrangements with Europe and Japan utilize one type of license, but developing countries need a different form. They feel that many countries in Latin America require a different approach from the others.

Most of the managerial aid that Clark provides is noncompensated and is usually provided to the licensee to help promote the success of the enterprise. For instance, in Mexico, Clark has a 26 percent joint venture wherein the company is involved in a noncompensated managerial contract with a company that provides substantial technical service on car transmissions. Clark's profit from the investment depends on its success; consequently, the company feels that such managerial responsibilities are justified, but the loss of such managerial expertise to international operations is still detrimental to Clark's internal group.

Thus, Clark feels that there could be a time when it would charge for the managerial services that accompany a know-how or technical disclosure fee. However, it appears that one must plan for these services and be able to provide them for a certain period of time. As it is now, Clark has provided managers (generally in noncompensated way) one or a few at a time—some even from retirement. So such provision has not been a great loss to the corporation internally.

Clark feels that there is a blurring of responsibility in a license contract where management is not mentioned. It is a far cleaner cut in such situations to note the managerial aspect. One "blurred" example is Clark's arrangement in Spain in which Clark builds lift trucks through a joint venture with a Spanish firm. Clark was the minority partner, providing their know-how and trademark, and had an export marketing agreement to market the output outside the host country's territory. During the recent depression in Spain, however, the firm ran into severe financial problems. Clark ended up having to send their financial people there. Thus, Clark provided a service not in the license agreement, but one that was deemed necessary to keep the project operating. Clark stated that generally, in one out of three situations, it supplies a service that Clark did not think would be necessary and that had not been included in the agreement. Clark received no compensation for helping in these instances, but the losses that might have been incurred otherwise would have been even greater.

Clark does not usually sell one-shot licenses or technology items. The company views licenses usually as a continuing document between two parties because Clark licenses "living products." It licenses by model but updates changes, usually called for in the contract, or provides amendments via new contracts. However, one official noted that an innovation might be to separate new licenses (transfer of technology) from management—that is, protect the license (know-how) via a paid contract on management.

Host. Bumar Union of Construction Machine Industry had a history of contacts with Western-based companies prior to its involvement with Clark Equipment. The first Western partner of the Huta Stalowa Wola combine (a factory complex in Poland) was the Jones Cranes Company, manufacturer of mobile cranes in the United Kingdom. This agreement, signed in 1966, is still in progress. Another agreement with a crane manufacturer, the Coles Cranes Company, also of the United Kingdom, was begun in 1971. Bumar also has contracts with Stetter of the Federal Republic of Germany for the manufacture of truck mixers, with the Swedish firm of Kockums for heavy duty transport trucks, and with the British firm of Hamworthy for hydraulic equipment. However, none of these agreements is as extensive as either of the agreements signed in 1972 with the two firms from the United States, International Harvester and Clark Equipment.

Axle production was not unfamiliar to Bumar prior to the agreement with Clark, but their axle design was considered obsolete, with the metal composition not adequate for extremely heavy vehicles. Thus, Bumar desired both an improved design from Clark as well as metallurgical techniques to improve the processing of the metal. Such improvements in their axles would enable Poland to produce vehicles with stronger construction—equipment vital to the country's plans for improvement and upgrading. In addition, such an axle would provide Poland with a viable export, either by itself or incorporated into a vehicle. The operation involving the axles was to comprise 400,000 square feet within the Bumar complex.

Transfer of Technology

The agreement between the Clark Equipment Company and the Bumar Foreign Trade Enterprises is actually composed of three agreements, each between Bumar and a separate entity of Clark. The Technical Assistance and Cooperation Agreement was signed by Clark's European subsidiary based in Switzerland, Clark Equipment A.G. (CEAG), while the Trademark Agreement was with Clark Equipment of Buchanan, Michigan. The third agreement, the Marketing Agreement, was between Bumar and Clark's marketing arm, Clark International Marketing S.A. (CIMSA). CEAG is Clark's representative for all licensing arrangements outside the United States, while CIMSA represents Clark in all international marketing responsibilities.

Turn-key. CEAG provided Bumar with a list of equipment drawn from Clark's own plants where axles were produced and assembled and assisted Bumar in their plant layout, but Clark did not build or equip the operation at Huta Stalowa Wola directly.

Information and Know-how. The know-how supplied by Clark included all technical information and production know-how necessary for the manufacture of the designated axles plus any relevant background knowledge related to the preparation of the materials, including metallurgical techniques. This knowledge regarding axles was up-to-date information, but it was also to be kept current, as Clark was to supply Bumar with any improvements or modifications made by Clark on its own production lines; Clark was not obligated to inform Bumar of any research being undertaken that had not already been implemented or actually put into production. This knowledge was transferred via documents and blueprints and through the training provided to Polish employees by Clark technicians both on the site in Poland and at Clark facilities in Buchanan and Jackson, Michigan.

The technical information provided by Clark included engineering drawings of details and assemblies; engineering bills of material; material specifications; assembly and production test information; inspection procedures; process sheets and route sheets; production, inspection, and assembly tool drawings and fixture drawings; metallurgical information; suggested machinery requirements and plant layout; SAE standards and AISI standards; and any available cross-reference information and patents. All information was to be in the language of the originating country (English, in this instance), and all conversion to the metric system was the responsibility of Bumar. The information was to be provided within 60 days after the receipt by Clark of the disclosure fee, or the required portions of it. Clark was to suggest which axles or portions of the axles should be initially purchased, using its best efforts to help in both the manufacturing and purchasing aspects.

Clark stated that the corporation felt that their partner was entitled to all the information that was necessary for that partner to build the product as well as Clark itself was able to build it. On the one hand, this does not mean that the partner needs to how how to design that product; but, on the other hand, that partner needs to know how to cut a gear tooth, how to heat tune it, and how to do all the manufacturing processes essential to that product's creation.

Both product and process technology were transferred. Clark emphasizes that the purchase price or disclosure fee is not just for blueprints or similar information but also for expertise, which is the aspect of technology transfer that is often the most crucial part.

Clark agreed to "use its best efforts to assist in purchasing or otherwise acquiring" parts that were not available in Poland or that were not to be immediately manufactured by Bumar.[3] This included parts manufactured by Clark subsidiaries and also extended to third-party suppliers. Furthermore, Clark was also to guide Bumar in the selection of manufacturing priorities related to axle parts.

Clark maintained the authority to approve the use of components being produced in Poland as well as to make certain that the parts were not used erroneously whenever certification of the construction or strength of the unit was at stake.

Management. No management as such was formally stated in the agreement, but Clark noted that whenever an extensive amount of technology is shifted, especially when information processing is included, management techniques are also passed along. Clark provided three types of packages, each carrying an implicit association with managerial expertise: all three agreements were to run concurrently and jointly expire, based on ten-year sales periods for each axle model.

1. The technical license, including the know-how;
2. The trademark agreement, which implies quality control measures;
3. The export marketing agreement, wherein all exports from this source are under Clark's control outside their home market area (for instance, Poland controls sales throughout the entire COMECON area, excluding the USSR, so Poland's home market is far larger than the country's boundaries alone).

Clark did not have any managerial commitments in Poland regarding the operation of the axle manufacturing plant. Such managerial control, Clark pointed out, is against the philosophy of Eastern Europe and legally not permitted, as no foreigner is allowed to give orders to one of the local residents. However, Clark was involved in training the Poles and such training included advice on plant operation, particularly as it related to efficient operation of the plant.

Clark's permission for the use of the corporate trademark was contingent upon three aspects: quality control, product design, and use of approved materials. The managerial area in such a situation might be best explained by noting the distinction between orders and suggestions. Clark did not order anyone at Bumar but only made requests.

Patents. All necessary patents were included with Clark's technology. Clark properly registered all essential patents in Poland, but Bumar, from that point on, had the responsibility of keeping the patent registrations up-to-date by paying the additional maintenance fees. Nonexclusive patent licenses were included in the agreement for both Clark and Bumar, with the additional stated obligation that Bumar disclose any improvements in the stated products back to Clark. Thus, even if Bumar did not want to take out a patent on an axle improvement, for example, they would be required to do so to protect Clark's rights.

Clark provided Bumar with approximately 25 patents (out of the 250 parts in each Clark axle) with the stipulation that Bumar not reveal this information to any third party for up to five years following the culmination of the Clark-Bumar accord.[4]

Trademarks. The Trademark Agreement between Bumar and Clark Equipment refers to all information provided by Clark Equipment A.G. through the Technical Assistance and Cooperation Agreement. Clark authorized the use of its trademark to Bumar not only within Poland but also throughout Eastern Europe (including Yugoslavia, Mongolia, the USSR, and India) for all approved axles produced through this agreement that had passed all designated inspections.

Bumar agreed to reimburse CEAG for the actual costs incurred in making all trademark tests, and all Clark products produced during the course of the agreement were to be trademark approved and were to carry the Clark U.S.A. designation. After the conclusion of the agreements, all use of the Clark trademark would cease except on items that had been produced prior to the completion of the agreement.

Clark provided Bumar with its trademark for all axles produced under the Clark technology, with the stipulation that periodic inspections be run to ascertain that all axles were meeting Clark standards and requirements to carry that trademark. Since Clark would be repurchasing some of the axles for its own use, the company was directly concerned that the axles be representative of corporate high quality. This buy-back provided another reason to have all modifications transferred to Poland as soon as feasible from Clark's other production facilities to ensure the interchangeability of all parts.

Clark will receive a royalty on all axles sold by Bumar for a ten-year period, commencing upon the first sale of the product. This royalty does not apply to all axles purchased back from Bumar by Clark directly.

Under terms of the warranties involved in this transfer, the parts produced in Poland must meet certain specifications. Should an item fail to meet those specifications, Bumar must then replace the item with one manufactured under Clark Equipment Company auspices in the United States, paying all shipping costs for the obtainment of the part as well.

Clark, through CEAG, maintained the right—but not the obligation—to be able to inspect the products being produced. Clark did agree to inspect and test at least one product from each different axle being produced to certify the use of its trademark.

CEAG agreed not to enter into an agreement of a similar nature with any other East European country for the duration of this agreement. Bumar was permitted to subcontract portions of the axle manufacture within Poland, but only after security precautions had been taken and Clark duly advised of the procedure.

Buy-back. The third part of the agreement between Clark and Bumar was the marketing commitment, in which Clark agreed to repurchase some of the axles produced by Bumar from Clark's technology for its own use at a price equal to Clark's own sale price from its factory in Buchanan, Michigan. Thus, Clark obtained axles already on the European continent for the same price as in the United States, but it avoided all transportation charges.

The Marketing Agreement was between Bumar and Clark International Marketing S.A., the marketing subsidiary of Clark Equipment that was based in Venezuela and is now based in the Caymen Islands. This agreement stated all price sources for parts, components, and full assemblies pertinent to the axle agreement, guaranteeing Bumar the availability, as far as possible, of all necessary parts, components, and spare parts. The Marketing Agreement also delineated portions of the world where CIMSA would use its best efforts to sell the axles produced by Bumar under the Clark trademark, for the joint interest of both parties. The agreement included the projected sales and production targets laid out on a year-by-year basis, starting from a year after the effective date of the contracts, and even stated the warranty that Bumar was to provide on the products resold to Clark. CIMSA was not held to any official number of axles to be repurchased from Bumar or sold under its marketing network. CIMSA was to provide its "best efforts" in working for such sales.[5] Clark was not to receive any royalty on any axles resold to CIMSA (at the ex-factory price, Buchanan, Michigan).

The buy-back arrangement is one that, ideally, should benefit both parties if the agent transferring the technology requires or desires an additional source of its product. In this way, the source is obtained without additional equity and the new producer is guaranteed a purchaser. No official buy-back level or limit, or starting date, was included in the Clark-Bumar contract, but Clark anticipated being able to buy products as soon as production was on schedule. The corporation did hope to be able to buy axles to help supply their French subsidiary—with the objective of covering major economic spheres throughout the world, particularly those within the European area.

The contract included an exclusive sales agreement but did not list exclusive territories. Bumar was permitted to market the Clark axles outside of its 11-nation territory if the axle was incorporated into a Polish-made vehicle.[6]

Responsibilities

Domestic Company. Clark's responsibilities in this project were to provide Bumar with all information and know-how relevant and necessary to completely manufacture the cast-housing plentary axles and to keep Bumar updated as to improvements on those axles for the duration of the ten-year production agreement. In addition, Clark was to provide suggestions as to equipment purchases and plant layout prior to the actual manufacturing process and to provide adequate training and consultative services so that Bumar employees eventually would be able to manufacture the axles without assistance from any Clark personnel. In addition to the technical assistance services, Clark also agreed to provide its trademark for the duration of the agreement (subject to adequate quality control measures) and to help in marketing the axles on a form of buy-back basis, since one of Clark's bases for undertaking this project was to obtain an additional source of these axles for their own marketing needs.

Host. Bumar's responsibilities were initially to provide suitable facilities in which to set up and operate the axle manufacturing equipment and to provide capable employees to be trained in such production. Bumar was to uphold all of its financial obligations to Clark throughout the course of the agreement, both as to the disclosure fees and the ongoing engineering services and royalty payments that were owed to Clark. It was to uphold product quality so as to maintain the Clark trademark. Bumar was also responsible for informing Clark of any improvements made in the axle by Bumar engineers, and was to financially keep all patents (Clark's and Bumar's) up-to-date, so as to protect the interests of both partners.

Corporate Perception of Benefits to Be Gained from Agreement

Clark equipment entered into the agreement with Poland for multiple reasons, but some of the most important ones were to obtain a source of supply without a financial commitment, to help offset prior development costs through fees paid for the technology, to maximize benefits from the technology before it became obsolete, and to capitalize on the opportunity to sell as many components or related parts as possible on an export basis. This last reason was generally aimed at temporary sales until the Bumar facility was on-stream and self-sufficient, but other sales were projected even after this point. Rather than prioritize these justifications for the agreement within this category, Clark officials felt that each reason was valuable and should attempt to be maximized. Bumar needed the axles, and Clark did not possess the supply capability either in the East or the West to supply Bumar without investing a significant dollar amount into production facilities. Thus, facing a supply problem where sufficient high-quality castings were not available, Bumar's invitation to establish a non-equity venture in Poland was a welcome and needed offer.

Secondary reasons for the agreement with Bumar in Poland included obtaining a toehold in the market, which would lead to greater involvement, and the sale of components that would be purchased to go into the product being assembled by Bumar. The possibility of the sale of additional products not directly related to the manufacturing venture with Bumar (except for service or replacement parts, which would be a direct spinoff) was also an underlying rationale and a hopeful one. These products might be desired by Poland or even other countries within the Eastern bloc as a result of the association between Clark and Bumar. Clark did not view the arrangement with Bumar as one that would help supply Clark with raw materials, although this possibility did exist.

Poland's first priority was to obtain an axle that would be a viable export commodity. Their desire for updated technology that would leapfrog the country's axle know-how into the modern era only seconded their primary concern with exports. The country also recognized a possible saving of hard currency because axles would be manufactured within the country and would no longer need to be imported. This justification ranks last, though, because Poland had been importing very few axles, since axles were manufactured within the country already.

Negotiations

The negotiations between Bumar and Clark were completed extremely rapidly, especially for such a first-time agreement. The reason behind such a rapid acquiescence is usually a strong desire for the agreement to be concluded; most often, this indicates a great demand for the product, which was the case in the Clark-Bumar agreement.

Parties Involved

The negotiations were conducted by Clark Equipment Company, U.S.A., but involved representatives of Clark Equipment A.G. and Clark International Marketing S.A. The agreement signed in 1972 was under the Bumar directorship of R. Mikoda.

Initiating Contact

Poland initiated the contact but Clark had been actively preparing to enlarge their European operations.

Length of Negotiations

The negotiating time between Clark and Bumar was unusually short. It was the first political agreement between Poland and a U.S. company and was completely negotiated in three months in Poland, start to finish, with ratification following within several months. Negotiations are usually assumed to require two years, with a two- to three-year time span considered standard.

Another factor affecting the speed with which the negotiations between Clark and Bumar were conducted was the fact that the arrangement was not only suitable to Poland, it also appeared to be ideally suited to Clark's immediate needs and desires. Clark had been "on the verge of pouring several million dollars into the expansion of its Belgian transmission plant when a fortuitous visit by the Polish government's commercial representative altered Clark's plans."[7] Poland was already building a new addition to its existing axle facility. Thus, Clark would be able to supply its own technology, suggest equipment that it favored, and purchase the axles it desired at a cost far lower than would be the case if imported to the United States. All these advantages would be obtained without any financial drain on the company—indeed, at a profit, due to the sale of the technology. Finally, unlike many West European arrangements, Clark would be able to maintain its access to its traditional markets in Western Europe and elsewhere while Poland concentrated on Eastern Europe, an area generally closed to Clark in any case. Furthermore, Clark was to receive a royalty payment for those axles sold by Bumar in Eastern Europe for the duration of the contract.

Clark also faced a complete lack of competition in this situation for, while axle producers form a limited oligopoly, it was found that the only other bid was ten times higher than Clark's. It is no wonder that the negotiations, which usually take several years, were completed within several months between Clark and Bumar.

Issues and Obstacles

The U.S. government did not disapprove of the Clark axle technology sale to Bumar in Poland, but Hayden notes that Clark preempted this possibility through a statement in their contract with the Poles that "absolved the company (Clark) of any obligation to disclose any technical data or information which . . . is primarily designed for or used in connection with military, naval, or other application relating to national defense or security." Hayden also noted that, in its export license application, Bumar committed the axles for use only only on nonmilitary vehicles and that, consequently, the license was immediately granted by the U.S. Office of Export Administration.[8]

English was the official legal language of the contract, as it is with all of Clark's legal agreements of this type.

Time Frame

Overview

The axle agreement was ratified on August 30, 1972, and is slated to run for ten years, during which time specified payments for the disclosure of the technology are to be paid on a regular basis. In addition, royalty payments are required based on plant production levels and are in effect for ten years from the time the product is first sold by Poland.

Special Considerations

The time frame of the agreement is ten years, from 1972 through 1982, for equipment installation and start-up of production. Production will run ten years for each model included in the contract, dating from its initial production in Poland. If an additional model should be added, this would extend the agreement ten years from that date.

The contract between Poland and Clark had no timetable or schedule, although certain estimates were included. The transfer and disclosure of information from Clark to Bumar began in December

1972, and two years were required before production was under way. It was anticipated that Bumar would be annually producing 15,000 axles and 3,000 cast-iron housings a year after 1976.

Financial Flows

Payments

The financial payments by Bumar to Clark were to be in two forms: an initial payment or series of payments for the technology itself, termed the disclosure fee, and a set of ongoing payments on an annual basis for the duration of the ten-year contract, termed the engineering fee and approximated royalty payments.

The majority of the disclosure fee was paid immediately following the ratification of the contract, with two smaller payments following in the next two years on the anniversary of the effective contract date. The percentage of the service fee due each year decreased at regular increments but a bottom limit was set. This minimum fee was stated for each year and increased, at regular increments, reflecting expectations for increased production and sales. All payments were in U.S. dollars, with no stipulation for inflationary increases apart from the current net price basis. No variable factor was tied to the minimum annual payments.

Payment schedules were arranged as such: Bumar would pay Clark the initial disclosure fee within 30 days of the signing of the contract, and Clark would provide the information to Bumar within one year from that effective date. For the duration of the contract, Bumar agreed to pay Clark a service fee and royalty payments determined by a quarterly report of Bumar to Clark. As mentioned before, Bumar was responsible for paying a certain minimum fee back to Clark each year, regardless of production rates. Clark's methods for determining the fees for a nonequity agreement vary from situation to situation. For example, the ongoing fee requested for the Bumar project was computed on a percentage of the U.S. sale price, with the decreasing percentage dropping from 4 to 3 to 2 percent over the ten-year period. In addition to making the royalty payments to Clark, Bumar agreed to pay Clark a 20 percent sales commission and was also required to pay a customs tariff of 11.4 percent on axles exported from Poland.

Purchases from the Domestic Company

Clark anticipated that some gain would be forthcoming from the sale of axle component parts to Poland in three ways: it would provide components until the manufacturing plants at Radomski and Stala

Wola were on-stream; it would enable Clark to sell components that would go into the axles being manufactured (more of a long-range prospect than the first area); and it would enable the company to sell exports to Poland that might be indirectly related to axle production but that were more a result of the Clark-Bumar relationship than the manufacturing venture.

Purchases by the Domestic Company

Clark expected to make few if any purchases of its own in regard to the agreement, apart from some equipment purchases that would be resold to Bumar, functioning as a middleman more than a selling agent. No Clark personnel were slated to be living in Poland for any length of time; the training schedule called for visits by Clark employees for no longer than three weeks at any one time. Thus, no purchases related to housing or accommodations were anticipated.

Personnel

Training Tasks

Domestic Company. Clark agreed to provide the necessary technicians and specialists for training in Poland for a certain number of days free of charge (with air fare and local expenses to be borne by Bumar, including a daily allowance). Any days over and above this stated number, however, were to be totally at the expense of Bumar.

Clark was to supply the technical personnel for the Bumar project, as needed, out of the wide resource of manpower available in the company. When metallurgists were needed, or heat treatment personnel, Clark was to send them to Poland. Clark put those domestic employees selected to work with Bumar under a special contract before dispatching them to Poland.

Hayden noted that Clark did anticipate some training needs, although the specific problems—worker indifference, careless craftsmanship, poor equipment—had apparently not been determined prior to the project.[9]

Host. The expenses of the Polish technicians brought to the United States were paid by Clark, but the anticipated costs of those visits were included in the initial disclosure fee.

Clark agreed to provide a certain number of man-days' worth of training to Bumar employees at Clark plants in the United States. For these visits, Clark agreed to pay hotel expenses and a daily stipend to each Polish trainee. (The expenses for this training were assumed to have been included in the price given for the initial disclosure fee.)

Protection

Leverage

The trademark was presumed to be a major point of leverage for Clark in dealing with the Poles. The Clark logo on the axles was considered virtually essential for Poland's axle exports. Indirectly, it was felt that the trademark provided the assurance that Bumar would be kept abreast of all developments or axle changes, since Clark would want all axles carrying its logo to be similar.[10]

However, the trademark was not considered Clark's only form of leverage. One reason why Clark appeared to have selected the nonequity form of technology transfer with few qualms was that the corporation felt it had a technology that was transferable step by step. It was pointed out that companies that must place all of their information up front are in a somewhat more precarious position with regard to collecting later fees or royalties. However, Clark felt, and still feels, that an equity position does not necessarily provide an adequate measure of control—even if that position is a majority one—as it was pointed out that "even if you own a subsidiary, you often do not control it." Hence, Clark has sought other forms of control through different leverage measures.

Inspection Privileges

Clark has found that the standards that the corporation requires for its operations abroad are sometimes actually stricter than the standards imposed domestically. However, the company's ultimate desire is to have total interchangeability of parts with no quality differential.

Patent/Trademark Defense

Clark offered assistance should any patent claims or liability claims result during the course of the agreement, but the company did not offer to directly defend any claims that arise. Another clause stipulated that Bumar was not to incorporate any of the information provided by Clark for use in other new or improved products without first obtaining Clark's approval or permission.

Updating improvements in the axle is a two-way venture between Clark and Bumar. Clark agreed to keep Bumar aware of all improvements in the axle, but Bumar also guaranteed to do the same for Clark and has granted Clark "nonexclusive rights to all (Polish) inventions, improvements and patents acquired or applied for" that involve the Clark axles during the course of the agreement. This area of protection or defense lasts only for the duration of the agreement.

Each different axle model manufactured by Bumar is covered by the terms of the overall agreement for a total of ten years from the time the first axle is sold, and is subject to payment of the royalties. However, after the ten-year period has expired for each model, Bumar is entitled to manufacture and sell the axles without the payment of further fees. However, use of the Clark trademark will not be permitted once the agreements have been concluded.

Legal Provisions

An arbitration clause is included in all Clark agreements; in the present case, the arbitration court in Switzerland is the designated one. A force majeure clause has also been included in all three agreements between Clark and Bumar.

In addition to the arbitration and force majeure clauses, Clark included a notation regarding the interference of "supervening powers" and a disclaimer regarding no disclosure of military information.

Project Assessment

Because the arrangements between Clark and Bumar were a first in several respects, Clark had no particular expectations beyond the aims of the agreement.

Expectations for the Future

One interesting aspect of the contract is that while the ultimate declared purpose is axle manufacturing, the possibility for joint production or development of metric axles or other projects of mutual benefit to both parties is stated on the first page of Clark's written agreement with Bumar.

International Harvester Company

Description

On June 7, 1972, International Harvester Company entered into an agreement with Bumar in Poland to manufacture crawler tractors or bulldozers at an existing facility in Poland. The enormity of this undertaking becomes more apparent when it is realized that this agreement included all aspects of the manufacturing and assembly process necessary to create this product. The ten-year agreement included rights to the International Harvester (IH) trademark and IH's obligation not only to furnish their most modern version of the crawler tractor but also to keep the facility updated as to improvements for the duration of the contract. The disclosure fee covering the initial transfer was not dependent on any buy-back, but research, engineering, and manufacturing fees (REMs) were limited to one-half the value of IH purchases from Bumar, either through the repurchase of components produced by the facility or the buy-back of the assembled final product.

Background

Domestic Company. Unlike some of the other companies involved in this study, IH had no prior involvements of the nonequity form, which it developed with Bumar in Poland or Raba in Hungary. In each of its 30 overseas facilities, it held at least a minority equity position. The Bumar project, with no equity contributed by IH, was a first for the company in over 140 years of operation.[11]

The undertaking in Poland was a completely new venture for IH in several regards. In addition to being the corporation's first nonequity venture, the plant in Poland was designed to manufacture crawler tractors, an item IH had never built on the European continent prior to this agreement. (Their plant in England, however, dating back to 1903, has manufactured the BT-20 model of the crawler tractor.)

The "international" emphasis in the IH name is appropriate, as a high percentage of IH earnings stem from international activities. In 1977 one-third of total sales of approximately $6 billion came from outside the United States.

A major internal corporate reorganization occurred in 1977, resulting in several relatively independent product groups.[12] In terms of the company's four product areas, $2.7 billion alone came from its truck sales, $2.3 billion from agricultural equipment, $731 million from Pay Line construction and industrial equipment, and $208 million from Solar turbo machinery. Since the reorganization at IH, the Pay Line group is now the division involved with the Bumar undertaking, even though all negotiations were initially conducted through the overseas division of IH.

Host. The agreement between Bumar and IH involved the Huta Stalowa Wola facilities, an industrial park complex. This plant already produced construction equipment, but the goal was to produce more updated versions that were both durable and modern. Bumar representative Systus Olesik stated that "the availability of the IH line of crawlers will complement our existing line of construction equipment for growing Eastern Europe markets and (Bumar) looks forward to being a source for International Harvester products in Europe and other markets best served by Poland."[13]

The Bumar Union of Construction Machinery Industry is the largest and most specialized manufacturer of construction machinery in Poland. It is the second largest manufacturer of construction machinery within the Council for Mutual Economic Assistance nations, and it ranks seventh largest worldwide in this type of manufacturing.[14] Bumar has been described as the manufacturer of nearly 90 percent of the construction equipment produced in Poland, with five major operations and 20 plants employing approximately 45,000 individuals.[15]

Managing director Jerzy Zaskurski attributed the rank and position achieved by Bumar to product specialization in heavy earthmoving machines and to a determination to produce basic hydraulic and drive assemblies, thus making it possible to "manufacture not only the finished machines but also all indispensible components."[16]

Bumar currently has its own steelmaking plant, steel and iron foundry, forge, and wire-drawing plant with numerous machinery and assembly plants within the Huta Stalowa Wola complex. Since 1967 Bumar has specialized in the manufacture and assembly of construction equipment. Its agreement with IH involves the manufacture of entire vehicles, with many of the pieces to be created at this facility, dependent upon the forging and casting skills of the Polish workers.[17]

The industrial combine, Huta Stalowa Wola, is Bumar's largest manufacturing facility and is located in the southeastern part of Poland. It originated in 1938 but was severely damaged during World War II. Since then it has been equipped and updated and today employs more than 25,000 workers in its 150-kilometer radius.[18]

Bumar has held cooperative agreements of various sorts with several companies around the world, including one with the Soviet Union for gearboxes and axles; International Harvester Company and Clark Equipment Company of the United States; Hamworthy, Jones Cranes, and Coles Cranes of Great Britain; Kockums of Sweden; and Stetter of the Federal Republic of Germany. However, the most extensive of any of these agreements is that with IH for the manufacture of crawler tractors, heavy loaders, and combustion engines—both as to designs currently in production and joint research and development on new products.[19]

It should be observed that despite the autonomy of each of the separate countries within the Eastern bloc, the Bumar Foreign Trade Enterprise has noted that it has a "good tradition of working closely with construction equipment producers in the USSR, Yugoslavia, Romania, and Czechoslovakia."[20]

Transfer of Technology

The initial cooperation agreement between the International Harvester Company and the Polish Industrial Combine Huta Stalowa Wola was signed on September 18, 1972, and included production, sales, and technical services for a series of medium-sized crawler tractors involving 140, 200, and 285 horsepower crawler loaders and pipelayers.[21]

Turn-key. This project was not a turn-key venture in the usual sense in that IH did not construct the facility or even design the layout for the equipment. Bumar did the layout but was able to seek the advice of IH for any difficulties that came along.

International Harvester did have "advisory obligations to Bumar in the matter of manufacturing process and technology, but no obligations or responsibility or the right to interfere in the selection of manufacturing equipment or organization of manufacturing," as this was entirely at the discretion of Bumar.[22]

Know-how. IH transferred all product technology for crawler tractors, vehicles that fall into the earth-moving equipment category. The transfer included all assembly line techniques and other process know-how.

It has been stated that under the 1972 agreement between Bumar and IH, "Bumar obtained the engineering and manufacturing know-how and the rights to manufacture and sell three basic sizes (five models) of crawler dozers and loaders: TD15C tractor dozer, 175C loader, TD20E tractor dozer, TD25C tractor dozer, and the TD25CS side-boom."[23]

Production know-how included the specifications required for metals and other materials going into the manufacture of the equipment. Information essential to this manufacturing process involved the tooling and machine techniques required to correctly operate the equipment and the inspections and specifications that must be met with regard to quality control measures. Process know-how generally involved assembly line skills but also overlapped into the area of machine operation.

The initial technology was to be updated by IH for the duration of the agreement until three years before its conclusion, which would be in 1979 of the ten-year period, or thereafter if the agreement had been extended. At that time, the information would become the property of Bumar if all other responsibilities had been fulfilled.

International Harvester agreed to provide all updates on the technology transferred under the agreement, but this pertained only to the immediate equipment included in the contracts. Thus, if IH were to develop a new machine that was substantially different from the existing ones, even if it qualified as a crawler tractor, this technology would not automatically be transferred to Poland. Or, if substantial changes were made in the existing equipment, such as a new axle assembly or another major part, this would not automatically be transferred either. In addition, the technology transferred by IH only pertained to those areas wherein IH possessed ownership. For example, the electric starter motors in the vehicles were purchased from third-party vendors by IH and thus certainly could not be transferred to Poland.

Management. No managerial aspects per se were contained in the contract in a separate area, but training know-how overlapped with managerial areas in the process techniques.

Patents. The crawler tractor agreement involved the major patent, which IH supplied.

Trademarks. IH supplied its trademark as one stipulation of the contract, but Bumar's responsibility in this regard was to produce a product meeting the strict standards imposed by IH.

Buy-back. While International Harvester "assumed no specific obligations to buy anything from Bumar" during the course of the project, as stated in their 1972 agreement,[24] it was anticipated that IH would purchase both components and entire vehicles back from Bumar, as reflected by the interconnection with the REM fee calculations.

The desirability of the buy-back stemmed from the lower cost advantage, but the fact that the vehicles would have been manufactured in a new plant, which was totally up-to-date and utilized the latest in equipment as well as in technology, would also be an enticement. Three French machine centers were purchased at a price of $1.5 million each. These were for manufacturing the crawler tractors. One of the four models being produced was to be of a completely new design, while the other three reflected models similar to those in current production.

Other Aspects. The contract between IH and Bumar provided Bumar with exclusive marketing rights to some countries, including all territory in Eastern Europe, and nonexclusive rights to other countries that were identified individually or by certain regions.

The exclusive and nonexclusive rights provided to Bumar by IH in the sale of the completed products related, in part, to the type of engine carried by the machine. Bumar was permitted to sell products carrying either the IH or its own Polish diesel engines throughout Poland and Eastern Europe. Outside of this region, the equipment must have an IH engine only if it is marketed through an International Harvester distributor in the nonexclusive countries.

International Harvester supplied all the requisite technology for the complete manufacture of the construction equipment but insisted that its own engines be installed in the machines, at least the ones distributed through its own international network of dealers. Furthermore, IH had negotiated a contract such that the Poles were committed to purchasing at least a certain amount of both completed products and components until their own facility was self-sufficient, thus guaranteeing IH of product sales that it would otherwise not have obtained.

Guarantees were in effect from the viewpoints of both parties in this agreement, in that Bumar was assured by IH that both the completed products and the components would be available to Bumar. This alleviated the possibility that Bumar would be caught short at some point before its own manufacturing facilities were functioning or be unable to complete its assemblies because certain components were unavailable.

Responsibilities

Domestic Company. International Harvester's contractual responsibilities to Bumar were to be available for consultation if desired on the renovation or installation of equipment for the manufacturing process, to provide complete product and process information and know-how, to make all necessary patents and IH trademarks available, and to buy back components and completed machines as required for resale through the international marketing network of International Harvester.

Host. The responsibilities of Bumar, in Poland, were to build and equip the facility—including casting and forging abilities as well as all manpower and raw materials meeting IH specifications—and to properly reimburse IH for its technology. Furthermore, Bumar had to make certain that the product produced, whether in component form or in its completely assembled state, conformed to rigid IH specifications before the IH trademark was applied.

Corporate Perception of Benefits to be Gained from Agreement

International Harvester's major reason for entering into this agreement involved the sale of whole vehicles (until the Polish facility produced its own) and continuous sales of components and replacement pieces. Since Poland already either manufactured or acquired from within CMEA tractors, wheel loaders, and pipelayers, in the conventional sense, the country did not present a viable market, as such, for International Harvester.[25] The opportunity, therefore, to sell completed products until the Polish undertaking was on-stream—and component and replacement parts even after that time—was considered to be the major benefit of the agreement. This increased market for whole units, component parts, and even service parts pertained not only to Poland itself but to the entire East European region. These sales involved components that were purchased for assembly into the machines produced by Bumar and for servicing the needs that arose throughout Poland that were not filled through conventional Polish channels.

The second reason for desiring the arrangement with Bumar was the direct profit gained through the technology sale itself. Since IH viewed the Eastern bloc as a closed territory for sales in its conventional manner, this technology sale provided a source of income that would otherwise not have been possible. Making use of this technology before it became obsolete was only a secondary consideration in this instance, since only the most modern technology was transferred to Poland. A more viable concern would have been that of the creation of a competitor should the supply of products exceed the demand. However, at the point in time that the agreement was negotiated the unfilled backlog was predicted to increase, and IH's concern was directed toward filling those orders, not worrying about competition.

The incentive to provide Bumar with the latest equipment designs was not limited to obtaining the contract. International Harvester stated that it wanted to uphold its own reputation as a leader in the area of construction equipment. This reputation would be enhanced through the additional attention paid to the Polish undertaking and the subsequent spread of its products throughout Poland and Eastern Europe.

IH's third reason for desiring the agreement was the potential source of completely assembled products and component parts. The buy-back aspect of the agreement was essential to obtaining the contract, but it also involved products that IH—through granting its trademark and applying the rigid qualifying measures that all products carrying its trademark had to meet—would be able to use within its own marketing plans.

International Harvester expected to benefit from the sale of products during its agreement with Bumar, regardless of the purchaser of the product. A royalty was guaranteed to IH from all products sold carrying its trademark, so all sales by Bumar would contribute to the sales of IH as well. Furthermore, all purchases by IH of either completely assembled products or components will be negotiated annually at a competitive price. Thus, IH will not be required to purchase products back at unfavorable prices.

The advantages to Poland and Bumar should also be examined. Bumar, through its agreement with IH, obtained the latest technology with regard to certain construction equipment. It not only obtained this equipment for its own internal use but also had a highly viable product for sale and export, thus providing an excellent source for obtaining hard currency.

Surprisingly, IH did not view their Polish operation as a toehold toward increasing involvement of an equity nature, despite the fact that IH had no prior nonequity arrangements anywhere else in the world. (Even though most Eastern European nations do not permit equity ventures, several countries in Eastern Europe had stressed their desire for equity joint ventures, even with Western-based companies.) Instead, IH viewed the Polish venture as one in which IH technology teamed with capital and manpower resources to produce the product. IH, ideally, gained an additional supply source through the capital and efforts of others.

Negotiations

Parties Involved

Negotiations between Bumar and IH were handled through the Overseas Division of International Harvester, the section of the corporation that had been responsible for all international activities from wholly owned subsidiaries to export licensing. IH's representatives included President and Chief Executive Officer Brooks McCormick, Executive Vice-President Omer G. Voss, and the president of the Construction Equipment Division, Jules C. Laegeler. Two representatives from the Overseas Division of the IH Special Operations Group were also valuable contributors: Robert J. McMenamin and William B. McIlvaine, Jr.[26]

Negotiations on the Polish side were conducted through the Ministry of Machine Industry. Bumar's own Foreign Trade Enterprise takes care of all imports and exports involving Bumar and is under the control of the general director of Bumar. Negotiators from Poland included the directors of the BUMAR Union, Huta Stalowa Wola, and the BUMAR Foreign Trade Enterprise.

Initiating Contact

Poland made the initial invitation to International Harvester because of the company's reputation in the construction equipment area.

Length of Negotiations

The contract between Bumar and International Harvester, signed in 1972, was not a hasty venture. Initial contact was made between IH and Poland in Chicago in 1968, and the first visits of corporate officials to Poland were made in 1969, three years prior to the signing.

Part of the reason for the extended period of time between the initial contact and the final contract approval was a delay caused by the Polish construction equipment industry's organization of its own foreign trade enterprise, BUMAR F.T.E., in 1971.[27] This foreign trade enterprise now has the responsibility for organizing the foreign relationships of the industry, both in export sales and licensing, with the ability to purchase foreign construction equipment for production in Poland as well as for domestic use.

The initial negotiations between IH and Bumar were the most difficult partly because both parties were venturing into new and unknown territory. Each time the tentative contract contents were revised or amendments added, a new fee schedule had to be drawn up. However, an accord was reached, and the success of the initial project—that of crawler tractors—was reflected in later agreements involving wheel loaders.

Issues and Obstacles

An initial issue facing the negotiations involved the nonequity nature of the Polish proposal. Since International Harvester had never ventured into such an undertaking before, this factor had to be worked out.

The large scale of the project was another factor adding to the length of the negotiations, because so many issues had to be considered. The fact that this was one of the few operations, especially in the industrial goods area, where total products were to be manufactured—not just components—expanded even more the extent of the undertaking and, hence, the negotiations. Such issues as the qualifications necessary for the IH trademark, the delineation of marketing territories, and the feasibility of the buy-back were involved. The buy-back issue was one of Bumar's requirements that IH was willing to consider in return for gaining access to the Eastern bloc.

The Poles were not willing to accept anything less than the most modern and up-to-date technology, since their intentions were to manufacture quality equipment that would benefit construction and other developments within their country, as well as to provide themselves with a viable and desirable product for sale and export. However, it would also be accurate to point out that International Harvester's concern with keeping its Polish partner up-to-date regarding the latest technology was doubly beneficial because of IH's plans to repurchase so many of the completed products, as well as components, itself.

Time Frame

The agreement between Bumar and IH was signed in 1972 and was to be effective for ten years for the manufacture and assembly of crawler tractors. The contract concentrated on the duties or responsibilities to be performed rather than specific deadlines.

Financial Flows

Payments

Bumar's payments for this technology were made in the form of an initial disclosure fee and subsequent research, engineering, and manufacturing fees (REMs) "expressed as a percentage of the value of units produced by Bumar during the term of the agreement."[28]

It was anticipated that financial flows would be high, as one statement from IH Executive Vice-President Omer G. Voss noted in a news release at the time of the signing in 1972: "Though it is difficult to accurately assess the full impact of this agreement on IH, we expect that it will improve sales by many millions of dollars annually. This venture will enable International crawlers and loaders for the first time to be manufactured and sold in substantial quantities in the rapidly expanding markets of the Comecon countries and will enable IH to increase its sales in Western Europe as well."[29]

International Harvester's fees from this nonequity transfer of crawler tractor technology were to come from several sources. The initial fee resulted from the disclosure payments. Furthermore, although the content of the completed machines was to be 50 to 60 percent Polish, IH was to supply the remaining components either from the United States or through its British subsidiary—an arrangement that provided another source of profit. In some cases, the sale was not just components but the complete machine. In other instances, it was replacement or service parts. In the latter instances, royalties were not paid if the sale was to Bumar. Other sources of profit were also generated in that IH would be able to purchase components back from the Polish facility for use in their own manufacturing operations in England or elsewhere. The Polish facility also would supply IH with complete machines at a negotiated price.

Purchases from the Domestic Company

By 1978 International Harvester expected to see an annual production of 1,500 completed units by Bumar with an invoice value of $100 million.

Purchases by the Domestic Company

IH anticipated no major purchases over and above the component or completed assemblies under the contract. No major housing or accommodation purchases were anticipated as the training was to be done in the United States with only periodic visits of IH representatives to Poland.

Personnel

Training Tasks

Domestic Company. International Harvester did not specifically indicate in the contract the number of individuals involved in the technology training. Their intention was to train the Polish supervisors in all aspects of the production and assembly of crawler tractors. The contract did provide for a specific number of man-months of training provided by IH here in the United States for Bumar employees. The contract also noted the number of man-months of IH personnel that IH would make available to Bumar in Poland; Bumar would cover their expenses. IH's approach was to have the Poles come to the United States at Bumar's expense for the training, so that the expenses on the part of the U.S. company would be minimized. International Harvester did expect some difficulty in getting corporate employees to accept positions involving frequent trips to Poland.

Host. It was recognized that extensive training would be required, but past experiences possessed by Bumar employees were taken as an indication that basic skills were already there and only needed to be applied to the particular IH manufacturing and assembly processes. Poland was to pay for all expenses for its employees, although IH was to provide Bumar trainees with a daily stipend while they were in the United States. The contract stated a certain number of man-hours to be contributed by IH representatives.

The technology was transferred via drawings and was also demonstrated in shop practices where utilization was stressed.

Protection

Leverage

International Harvester was not particularly concerned about the possibility that Bumar would not provide the appropriate royalty payments, based on Bumar's sales of machinery and components. This lack of concern was also based on the excellent working relationship between the two parties and on the IH trademark. Should Bumar not honor its agreement with IH, the use of the trademark would no longer be permitted. Hayden pointed out that "the Poles know that without the quality guarantee assured by a U.S. company's trademark, their own products would have little chance of competing in Western markets; without access to the U.S. company's product improvements and its West European servicing agents, their own output would not long endure in the competitive West."[30]

Domestic Company. Thus, use of the trademark constituted a considerable form of control. The threat of losing the IH trademark was expected to provide IH with substantial leverage, since the company believed the trademark's loss would have a detrimental impact on Bumar's operation.

Another form of leverage was contained in the arrangement: Bumar must purchase engines from IH to qualify for the IH trademark. It was felt that Polish-designed engines in U.S.-designed machines would not provide a high enough level of performance. Thus, should IH refuse to sell engines to Bumar, the entire production capability would be limited to selling only components and complete vehicles utilizing the Huta Stalowa Wola (HSW) trademark. Therefore, the components supplied by IH, in addition to the trademark, also constituted considerable control over the status of the project, even though the technology itself had already been transferred.

International Harvester also stipulated that the technology provided to Bumar not be transferred to anyone else. At the conclusion of the agreement—either after ten years or whenever the agreement was to end if extensions had been applied—Bumar was to have ownership of the technology if all responsibilities and guarantees had been met. This technology would include all updated aspects developed throughout the course of the agreement, up to three years prior to the conclusion of the agreement.

A technology guarantee was provided by IH to the effect that Bumar would be able to create vehicles meeting IH standards if all instructions were correctly followed and the proper raw materials were used.

Host. Bumar also had a high degree of control in this operation, in part because IH requested that the products being produced at the Polish facility be available to supply IH's own subsidiary in England. Thus, IH's assembly plant in England needed the products being produced in Poland, while Poland also required some of the components from IH.

The only financial guarantee posted throughout the operation appeared to be the letter of credit posted by Bumar to cover the first payment of the disclosure fee. This was a onetime event, however, and was not expected to be repeated.

Inspection Privileges

Inspection provisions were included for IH because of the strict standard requirements accompanying the IH trademark. IH was guaranteed access to inspect and test the workmanship of all products being produced under its trademark.

Patent/Trademark Defense

Registration in Poland of patents involved with the technology was required. The use of the IH trademark was open to Bumar, as long as the stipulated rules and regulations were followed.

Legal Provisions

An arbitration clause was included in the contract, as well as a force majeure notation.

Project Assessment

Factors Expected to Make the Project Easier than Usual

The past experience of the Bumar employees was considered a great advantage because so many of the procedures would already be familiar to them. International Harvester also felt that the training sessions conducted in the United States, where similar production was actually taking place, would be an ideal way to provide a working example for the Poles. The past experiences of International Harvester throughout the world were also expected to be of great help, because although none of the arrangements had been of a nonequity nature, the international background had provided IH with a broad scope of potential problems and their plausible solutions. International Harvester felt that it entered into the agreements with Poland with general expectations but no fixed, firm ones.

Factors Expected to Make the Project More Difficult than Usual

Several aspects of this technology transfer appeared most formidable. First, this technology sale included total vehicles. Unlike other transfers that involve only a portion of the vehicle, such as the axle or braking components, IH agreed to transfer the entire knowledge behind one of its more desirable units. Furthermore, it guaranteed that through the correct application of its blueprints and training these vehicles would be able to be manufactured to meet specifications that would permit the use of the IH trademark.

The buy-back aspect of the agreement provided further pressure on IH to succeed in this endeavor, since the corporation's hopes for an additional product source partly depended on the successful production of the intended machines.

Another difficulty involved the fact that this arrangement was a first with regard to a transfer of this type with Eastern Europe, a region somewhat unknown to IH. This factor, combined with the totally unfamiliar nonequity type of contract, required IH to feel its way as it went along. Due to the nature of contracts and agreements with Eastern Europe, the transfer had to be totally laid out prior to the start of the agreement, thus forcing IH to project its expectations without much past experience to go by. However, IH felt totally confident that the project would succeed and also felt that it would be able to work through any problems or difficulties that might arise.

Expectations for the Future

International Harvester hoped that the agreement with Poland would result in (relatively) immediate product purchases by Bumar from IH until the facility in Poland was producing on its own. In the long run, IH hoped that this initial contact with Poland would lead to further agreements both in Poland and throughout Eastern Europe. However, these expectations, while important, were secondary to the hopes that the immediate project would be fruitful and profitable, particularly in terms of the technology fees and equipment sales.

Honeywell

Description

The 1974 agreement between Honeywell, Incorporated in the United States and Metronex/Mera in Poland involved the class 41 transmitter, an electronic pressure transmitter based on a piezoelectric silicon sensor. Technology developed in the United States was to be transferred into Polish production with the help of only the U.S.-based division. The agreement was scheduled to last five years but did have provisions for extensions.

Background

Domestic Company. The Polish government insisted that the agreement include the sensor technology, specifying that all information be the latest and most up-to-date available. Surprisingly—considering Honeywell's usual reluctance to provide such current technological information—the company agreed to this demand. This was unusual on two counts. First, Honeywell controlled the only source of this sensor technology. Second, the stated interest of the corporation has been to sell products, not the technology behind the products. Still, an accord was reached and the agreement was concluded—subject, however, to the approval of the U.S. government for the transmittal of sensitive technology.

Hayden reported that Honeywell was more inclined to sell its technology to a developing nation than a developed one, because it felt that the developing country would be less likely to develop into a competitor.[31] Eastern Europe falls somewhere between the developed and the developing countries. Honeywell was concerned that their technology—especially that relating to the valuable sensor assembly and construction—not fall into the wrong hands, which included the Soviet Union. The company also felt, however, that Poland was particularly desirous of working with them at least for the immediate future and, consequently, would not be willing to risk that relationship by transmitting sensitive technology to others.

This contract provides an example of a technology exchange utilizing a product available only through the Honeywell auspices. The corporation categorizes its operations into four areas: environmental systems and controls, industrial systems and controls, aerospace and defense, and information systems. Its total annual revenues are close to $3 billion, with information systems providing more than $1 billion alone. Honeywell has transferred technology in a manner similar to the Metronex arrangement to other parties in Holland, Italy, and Japan from the 1940s through the 1970s. Various Honeywell products have been involved, such as residential heating, industrial products, and aircraft control devices and systems.

The corporation is currently involved with a technology transfer agreement with Poland through a license involving the class 41 transmitter developed by the process control division located at Fort Washington, Pennsylvania. This technology transfer to Poland is under the industrial systems and controls area, which provides control devices, analog- and microprocessor-based instruments, and computer-based systems for data acquisition, monitoring, and control of industrial and electric utility processes.

Host. The agreement was signed by the Polish Foreign Trade Enterprise, Metronex, which is part of the Polish trade union, Mera. Metronex is an import/export agency in Poland for a particular category of goods under the umbrella-like auspices of Mera.

The Technical Assistance Agreement between Honeywell and the Polish Foreign Trade Enterprise, Metronex, effective December 19, 1974, was not the first contract between the corporation and Poland, although it was the first to have totally U.S.-based involvement. Three license agreements had been established in 1973 between Poland's purchasing enterprise, Metronex, and three European-based subsidiaries of Honeywell. All three agreements involved some aspect of technology exchange. The Honeywell Europe S.A.-Mera/Metronex Agreement of May 10, 1973, (also called the Milan Agreement) involved a license for Poland to manufacture two devices—the differential pressure transmitter no. 411 and the pressure transmitter no. 412—for five years. The Honeywell GmbH Agreement, signed in June 1973, involved the manufacture of the vutronik line of valves to control fluids in large pipes. Honeywell's Scottish subsidiary, Honeywell, Limited, licensed production of the vutronik line of recorders and controllers in November 1973.

Transmitters are important to Poland because of their versatility, especially in the area of petroleum extracting and refining. Since Poland's plans for the future include a heavy emphasis in this area, this technology is of prime importance to the country's chemical industries. Hayden noted Poland's intention to double its current oil consumption by 1980, and one-third of the country's 1976-80 expenditures were assigned for the construction of new refineries and petrochemical plants.[32] These transmitters are used as monitors on pipes carrying liquids. Solid state construction makes the item both durable and unlikely to break down. It also provides an extremely accurate gauge of the pressure of the liquid within the pipe. Thus, the Honeywell product is in demand both because of its design and also because of the increased need for such a product.

Transfer of Technology

To efficiently transfer this new technology to Metronex, Honeywell accepted responsibility in four areas. First, Honeywell agreed to transfer to Metronex in Poland all testing equipment required to test various types of class 41 transmitters. Second, it also agreed to provide Metronex in Poland with class 41 transmitter acceptance kits. Third, utilizing Honeywell's employee services, the company would train Metronex specialists at Honeywell facilities in the United States (Fort Washington facility). In addition, Honeywell supervisors would be sent to Metronex facilities in Poland to direct the Metronex staff in assembling the class 41 transmitter acceptance kits into class 41 transmitters, using special tests and testing equipment. Fourth, Honeywell pledged to support Metronex specialists in the United States and Poland regarding the manufacture and utilization of sensor assembly.

Turn-key. Honeywell's involvement in the operation was most comprehensive but, since Honeywell did not construct the building, this project does not fall under the usual definition of a total turn-key operation. In this instance, an existing facility was remodeled. Honeywell did, however, supervise the entire equipping of the plant, right down to the screwdrivers. This included providing equipment either constructed at Honeywell's operations in the United States or purchased from third-party suppliers through Honeywell's auspices in the United States. In addition to the purchase and installation of the equipment, Honeywell supervised the tooling of equipment when necessary. This equipment came under the categories of device acceptance kits, special testing equipment, and certain production tools including tooling facilities. Upon expiration of the contract, these items will remain the property of Metronex.

Information and Know-how. All product and process technologies initially provided pertained to the class 41 transmitter and included all information related to its construction and testing—with the exception of the sensor technology, which would be held in abeyance until the fifth year of the agreement. Honeywell agreed to sell to Metronex the special testing equipment required to monitor the manufactured assemblies and the acceptance kits. This was considered part of the total Honeywell know-how package, encompassing the manufacturing, testing, servicing, and repairing of the class 41 transmitter assemblies or subassemblies. Broken down somewhat, this package included the testing (but not the manufacture, servicing, or repair) of electronic components and subassemblies of these transmitters that are also used in other Honeywell products; second, the testing (but not the manufacture) of the portions of these transmitters that are purchased from third-party sources; and third, the testing and servicing (but not the manufacture) of machines and other apparatus purchased by Honeywell from third-party sources and used to manufacture the transmitters.

The class 41 transmitter technology transferred included the class 41 pressure, differential pressure, flange-mounted level, extended diaphragm flange-mounted level, absolute pressure, and compound transmitter models. In addition, Honeywell agreed to transfer all know-how relating to developments and improvements of these class 41 transmitters during the contract period.

This agreement included all information pertaining to the class 41 transmitters that was owned by Honeywell at any time during the term—excluding items involved with the sensor assemblies from third-party sources. Honeywell sensor assembly know-how was considered a special aspect of the technology transfer and was the subject of some special provisions.

The English language version of the agreement was the official version. All know-how documentation would be in the English language.

Management. Management of the plant was not the responsibility of Honeywell, although training in managerial and production techniques was provided both on site in Poland and in the United States at the Fort Washington plant.

Patents. Honeywell agreed to furnish patent information to Poland should the Polish patent rights be challenged by a third party. Honeywell agreed to provide Poland with a list of all patents owned by Honeywell related to the class 41 transmitter within three months of the start of the agreement, with an updated listing one year before the agreement's conclusion.

Trademarks. The agreement provided patent rights and trademark rights of a sort. The label of the Polish product is to read "Made in Poland-MeraPnefal under license from Honeywell, Inc." This notation must cease when the agreement expires.

Buy-back. The agreement did contain an allusion to a buy-back clause, but only on the condition that the items selected be of comparable price and quality in meeting recognized needs. It contained no "hold-back" clauses, referring to withholding of payment, products, or services, should responsibilities of either party not be met.

During the five-year period of the agreement Poland will be allowed to sell the completed assemblies within the Eastern European area, including Poland, Hungary, Czechoslovakia, Romania, the Soviet Union, Bulgaria, East Germany, Yugoslavia, Albania, and Mongolia. They will also be able to sell the assemblies outside the Eastern European region after the five-year agreement is concluded.

The implication that Honeywell would buy back an equivalent product if desired quality and need existed suggests the pressure that was placed on the company to include the buy-back aspect in the agreement. The fact that Honeywell had the leverage aspect of proprietary claim to the product, however, appeared to be instrumental in being able to avoid a total buy-back commitment.

Other Aspects. Total assemblies and subassemblies would also be provided by Honeywell to help meet production quotas until the plant was operating on its own and also to help train the Polish workers in the assembly process.

Responsibilities

Domestic Company. Honeywell's responsibilities, in general, were to equip the factory, train the workers, transfer know-how regarding the class 41 transmitter, and provide certain testing apparatus. Their duties were not to end until 75 units, which met all specifications, came off the assembly line. However, this responsibility did not include overall management of the project, only specific approval of machines and tooling levels. Honeywell was responsible up to the point that the host could manufacture the instruments within certain specifications.

Honeywell agreed to provide the services of its company specialists and supervisors to perform specified training tasks and to provide special testing equipment and device acceptance kits. Specified documents of Honeywell know-how would be provided by certain dates, particularly, but not limited to, the first six months of 1975 or within certain time periods following the release of manufactured items. Documentation refers both to the current documents for the class 41 transmitter and to all future class 41 transmitter changes or revisions made by Honeywell.

Honeywell also was to furnish samples of the transmitter or portions of the units involved in the creation of the transmitter at specified times and agreed to test five transmitters each year for the duration of the agreement. Honeywell would also supply additional employee services should Metronex request such services.

Honeywell agreed to make available the sensor assembly know-how, including employee services, during calendar year 1979. Only the list of machines (including manufacturers utilized in the sensor assembly) was previously provided in early 1977.

Access to the sensor technology would appear to be more of a beginning than an end to the association between the two parties because the manufacturing capabilities required to produce this intricate item are quite rigorous. Thus, should Poland decide to go ahead with the manufacture of the sensor, expensive tooling of the equipment and additional training of employees will be required and would be subject to further agreements and an extended time limit.

Host. A summary of Metronex's responsibilities under this agreement would be that it was to follow all instructions specifically as laid out by Honeywell, and it was not to sell or provide services connected with the class 41 transmitter outside the Eastern European territory until the specified time limit had been concluded. In terms of its own aims, Metronex hoped to eventually manufacture a product with 15 percent U.S. input, or less, that also met all specifications. In addition, the Polish company was to inform Honeywell of any improvements made by Poland in the class 41 transmitter or related process control instruments, including all related technical information. This could potentially lead into a joint development venture, should they improve on the product for either local or international production.

Corporate Perception to Be Gained from Agreement

Honeywell entered into this agreement with Poland hoping both to increase the corporation's penetration into Eastern Europe in general and to sell more products. Hopefully, the Honeywell name would become better known throughout Poland and Eastern Europe so that items not even indirectly associated with the class 41 transmitter would also be purchased from Honeywell. Honeywell did not view the manufacturing situation as a source for obtaining additional transmitters. Thus, the sale of technology was the main initial motivation for Honeywell in desiring this agreement with Poland. Accompanying benefits included the sale of assemblies or subassemblies (particularly until the Polish production capabilities were functioning) and the access gained to a new market.

Negotiations

Parties Involved

Domestic Company. A three-man team was involved in the contract negotiations: one person was a representative of the Minneapolis-based headquarters; one was from the Fort Washington, Pennsylvania, manufacturing plant where the class 41 transmitter is produced in the United States; and one was an experienced negotiator highly familiar with the Polish culture, language, and previous Honeywell European agreements.

Host. The agreement was made between Honeywell and Metronex, a part of the Mera Union within Poland. The individual factory employees in Poland were only generally consulted during the negotiations, although the Fort Washington plant's representative was an integral part of the Honeywell team. Implementation of the agreement would be the responsibility of the two respective plants, however, both as to equipping and training.

Initiating Contact

The Polish government was interested in obtaining access to Honeywell technology and made the initial overtures, but most of the contact grew as an extension of the discussions carried over from the European agreements signed in 1973.

Length of Negotiations

Total time involved in the negotiations, including discussions on the sensor technology, took approximately one and one-half years. Much of this time was spent obtaining permission, and then discussing the methodology, for the transfer of the silicon sensor technology, the particularly desirable aspect of the class 41 transmitter.

Issues and Obstacles

Security and Approval. U.S. government approval had not been secured by the date of the contract signing in 1974. As a compromise, Honeywell agreed to provide all other information initially and to offer unlimited purchases of the sensor for the course of the agreement. During the final year of the agreement in 1979, Honeywell would then transfer the sensor technology if all permission had been granted. Permission to provide the sensor data was eventually granted by the U.S. Department of Commerce as of January 1979.

The negotiations were held both in Poland and in the United States. However, the location selected as the depository for the sensor data until the 1979 deadline was in a third country considered neutral by both parties. The appropriate documents were placed on reserve at the Swiss Depository Bank in Geneva during the first six months of the agreement.

To ensure that the technology would be supplied to the Eastern European recipient at the appropriate time, regardless of political or other intervening forces, the data placed in the bank on microfilm carried with it instructions that it was to be turned over to the appropriate Polish authorities in January 1979 should unforeseen circumstances otherwise prevent the final transfer to Mera. The only condition that would nullify these instructions would be default by the Poles on the terms of the agreement when, despite notification of the default, the situation had not been corrected. Barring war or other circumstances beyond the control of either party involved, the turnover of the documentation would include updated techniques based on whatever improvements were made over the course of the five years of the agreement.

Honeywell's corporate philosophy is to sell products, not necessarily technology. Foxboro, a U.S. competitor of Honeywell in the process control division, was also being considered for this Polish contract, although the Foxboro products were different in composition from Honeywell's. Hayden states that part of the reason for the acceptance of the Honeywell contract was that company's willingness to sell its technology for the class 41 transmitters and to provide the training necessary for their manufacture. The novelty of such willingness was the selling point that provided the basis for the acceptance of their offer over all the others.[33]

Detailed Contract Format. The format of the contract was extremely detailed, designed that way partly because of Honeywell's experiences with its European subsidiaries. It required strict adherence on the part of the Polish engineers to each aspect of the transfer; screws were not even to be changed without permission. This was viewed as essential by Honeywell. Past experiences, particularly with its Scottish subsidiary, had indicated that comprehensive contracts produced better actual operation. Thus, despite the desire on the part of the corporation to conclude the negotiations as quickly as possible to free its negotiators for other matters, each element of the transfer was painstakingly explored and delineated.

Time Frame

Overview

The Technical Assistance Agreement between Honeywell and Metronex, effective as of December 19, 1974, was to run for a period of five years, until December 31, 1979. The agreement could be extended, subject to terms mutually acceptable to both parties, if such a request was received 12 months prior to December 1979.

Within the project's five-year time frame, one to one and one-half years was allowed for start-up of the assembly, or from the opening on October 19, 1974, to anywhere between December 1975 and June 1976. It was estimated that four months would be necessary for equipment installation, and an additional two months for training equipment operators. The entire agreement was to conclude December 31, 1979, unless an extension was requested.

Special Considerations

Know-how. Honeywell listed the documents that contained all the know-how regarding the class 41 transmitters and provided Metronex with one copy of each document during the six-month time period extending from January 1 through July 31, 1975. If any class 41 transmitters were developed during the period of the agreement, the contract required that the necessary documentation be sent to Metronex within 30 days after Honeywell approved the item for manufacture. At the conclusion of the agreement, Honeywell was to provide a transparency and a microfilm copy of all drawings involved with the know-how package (in metric dimensions) as well as a copy of all Engineering Change Orders relating to the transmitters and subsequent revisions.

The transfer of the sensor assemblies, subsequent to approval by the U.S. government, was to begin January 1, 1979, with all information transferred by February 1, 1979. For security purposes, the information being transferred had to be in the Swiss Depositor's Bank within the first six months of the start of the agreement, or by June 1975. Prior to the ultimate transfer of sensor assemblies, Honeywell would provide Metronex with a complete list of all machines and other apparatus used to manufacture the sensor within 30 days after January 1, 1977.

Payments from Metronex to Honeywell. Payments were spread out across the five years of the agreement, although the greatest bulk of the money was to be paid within the first two years in the first three payments. With the exception of the first payment, due shortly after the signing of the agreement (late in December 1974), all payments were to be preceded by a letter of credit that was to be lodged with the Philadelphia National Bank by a certain date.

Products Purchased from Honeywell by Metronex. Honeywell agreed to sell Metronex both full units of the class 41 transmitters and parts and subassemblies in order to meet needs for the product until the Polish manufacturing facility could get under way. The sale of parts and subassemblies would provide a training ground for the assembly of the units.

Prices for the respective units or subunits were listed in the agreement with a full price given for full units (including a 10 percent discount off the current retail price) and a stated price through the end of 1975 for subunits, with a pricing formula to be effective after that period of time.

A fixed price was applied to all sensor assemblies purchased through 1977 (with a maximum of 2,000 units per calendar year); a pricing formula would apply from that time through 1984. Metronex was also supplied with a procedure to follow for projecting future needs for the products.

Patent Listing. An initial list of all patents owned by Honeywell and related to the class 41 transmitters were to be passed on to Metronex within the first quarter of 1975. An update of this list was to be supplied by the end of 1978.

Training. The agreement included a training schedule that provided for the man-days to be supplied by Honeywell and Metronex as well as a complete list of the types of training to be offered and the description of the skill areas or backgrounds of the persons receiving such training.

Production Deadlines. The project was intended to go into production within the first quarter of 1976 after the following conditions had been met:

1. Polish operators selected for production
2. All special testing equipment installed
3. Production methods decided on
4. All class 41 transmitter acceptance kits ready
5. 75 transmitters meeting specifications ready to be assembled

Extensions. The agreement was open to extension if application was made 12 months prior to its conclusion in December 1979 (or, by December 1978). The extension would not affect the transfer of the sensor technology; and, while any extension would allow Metronex access to updated technology in the class 41 transmitter area generally, an update on the sensor technology would not automatically be a condition of the extension.

Honeywell's attitude toward an extension of the agreement appears to be quite positive, because an extension would mean more product purchases. Poland's desire to continue the agreement would provide the country with more technology updates.

Financial Flows

Payments

Prior payments to Honeywell had been in the currency utilized by the subsidiary. The Scottish subsidiary's agreement, for example, specified English pounds, while the U.S. corporation's agreement of 1974 requested dollars. All payments from Metronex were to be in U.S. dollars, net of any taxes or other charges by the government of Poland.

Honeywell was to assume all responsibility for arranging and paying for any export licenses required for this agreement. Metronex was to arrange and pay for all required import licenses and/or duties.

The payments were divided into two general categories, the first constituting a disclosure fee for the technology and the second comprising subsequent payments reflective of the products being produced (although there was no percentage listed or specific royalty-type payment schedule established).

All payments were monetary in nature; no products were taken in exchange. Setting the appropriate price proved difficult because there were no Polish products of comparable price or quality. Since the entire financial basis was specified in the contract at the beginning, the wrong price could well have jeopardized the profitability of the entire venture.

Honeywell's compensation for the transfer included an 8 percent annual escalation fee applied to subsequent payments after the initial disclosure fees. These subsequent payments were termed the update fees because improvements in the product would be provided to Poland for the duration of the contract.

The payment schedule specified in the agreement included the amount of the payment, the date it was to be received, and the date of the letter of credit that was to precede the payment. The first payment, due in late December 1974 at the start of the agreement, was the only one that did not require a letter of credit.

For all services and know-how that Honeywell would provide, Metronex agreed to pay Honeywell through the Philadelphia National Bank according to the predetermined payment schedule. Additional Honeywell employee services that might be requested by Metronex over and above the amount scheduled in the contract were to be reimbursed within 30 days after Metronex received the monthly invoice from Honeywell for such services.

Purchases from the Domestic Company

Honeywell agreed to sell complete units to Metronex until the Polish manufacturing facility was in operation. The price for such units included a discount from the then current Honeywell list price, and the items were guaranteed for 18 months from the date of delivery. Delivery was to be within 20 weeks of Honeywell's receipt of the order.

Honeywell further agreed to sell parts and subassemblies of the transmitters to Poland through the course of the agreement (and, in the case of the sensor assemblies, until December 31, 1984). The agreement specified that the ex-factory price for these units would be effective up to December 31, 1975, and thereafter would be computed according to a formula that included an escalation factor to take rising costs into account.

Sensor assembly prices were set through December 31, 1977 (a total of 2,000 units for each calendar year); prices were then computed on the basis of another formula for the period 1978 through 1984. Both formulas took various cost-of-living index factors into account in determining an equitable price for the items. Honeywell guaranteed these items for 12 months from the date of delivery. Delivery was to be within 18 weeks of Honeywell's receipt of the order for subassemblies and 14 weeks for parts.

Eastern Europe generally does not manufacture many spare or replacement parts. In this agreement, Honeywell insisted that Metronex be willing to use up to 5 percent of its extra component parts to rework any items that did not meet intermediate or final inspection. Honeywell would supply any additional components needed beyond this amount, free, unless the shortage was due to carelessness on the part of Metronex.

Payments for both the complete units and the parts or subassemblies were to be via a letter of credit through the Philadelphia National Bank, submitted 30 days prior to the shipment of any items. All payments were to be completed before any products or equipment were shipped. Should delays occur on the part of Honeywell, a change, based on a percentage of the price, would be assessed.

Purchases by the Domestic Company

It was anticipated that no purchases would have to be made by Honeywell apart from equipment supplied to Metronex under the agreement.

Personnel

Training Tasks

Domestic Company. The agreement laid out the daily charges for Honeywell production specialists in Poland, computing the actual travel and living costs plus a daily charge that increased slightly for each year of the agreement. The agreement also listed the daily charge for each Metronex employee at the Fort Washington facility, which also increased slightly each year of the agreement. Honeywell agreed to send to Poland one production engineer for six weeks (30 man-days) and one quality/test engineer for three weeks (15 man-days); together they would account for 45 of the 60 man-days projected for 1976.

There was a separate, higher cost structure for the 1979 transfer of sensor technology for both the Honeywell specialists in Poland and the Metronex employees in the United States (in this case, in the factory in Minneapolis, Minnesota) that reflected the more complicated technology involved in the sensor's production.

Costs associated with training and traveling were absorbed by the respective entities. When the Poles came to the United States for training sessions, they paid for their travel and living expenses. Similarly, when Honeywell employees traveled to Poland, the corporation paid for those expenses, since the negotiated fees paid to Honeywell under the agreement had included a portion for the travel and training expenses involved in the transfer. Should the training sessions in Poland requiring the presence of Honeywell employees exceed the limit in the contract, Poland would be financially responsible for those extra trips or additional days.

Host. In the case of these manufactured transmitters, training was considered particularly important because of the accuracy needed for product integrity. The training process supplied by Honeywell's employee services to Polish personnel in the United States began with a statement of objectives for each area of training, the specific topics to be covered, the time required, and the Polish personnel positions or qualifications selected to receive the particular training. These experience or position areas on the part of the Polish delegates included the positions of quality engineer, materials engineer, design engineer, production engineer, and foreman.

The agreement identified ten skill or task areas that would be taught to Polish personnel in the United States: receiving inspection, subassembly production, test operations, printed circuit board assembly, troubleshot operations, theory of operation, quality assurance, materials control, documentation, and final assembly operations. Polish engineers would first be brought to the United States for training, then would continue that training back in Poland as others were trained at the same time. Updating would also be provided over the course of the agreement either in Poland or in the United States.

Protection

Leverage

Domestic Company. Honeywell maintained a high degree of control over the operation in several respects. First, because of the upcoming transfer of the sensor technology, Poland had an additional incentive to maintain good relations with the firm. (This is not to say, however, that amiable relations would not have been present otherwise.) Second, Poland intended to purchase sensor assemblies from Honeywell prior to the sensor technology transfer. These assemblies formed an integral part of the class 41 transmitter, and thus, the purchases were essential to the operation of the item. Third, there was a high possibility that Poland would continue to purchase these sensor assemblies even after the transfer of the technology, due to the extremely high cost of tooling involved in equipping the plant for sensor manufacturing. Fourth, Honeywell was essentially the sole source of this item or product so that, should the relationship break down, the Poles would have no other means of obtaining the items. Because these items were necessary—or at least greatly desired by Polish industry—there was a high incentive to make the entire agreement successful. Fifth, since the contract involved updating, the Poles had more to gain through maintaining relations for the duration of the contract.

A secondary form of control rested on the Poles' ability to learn the new technology. If it appeared that the Poles were not capable of mastering this manufacturing operation, Honeywell had the option of noting this and subsequently stopping the transfer.

Honeywell guaranteed in writing that its responsibilities for successfully transferring the technology—in terms of product know-how and process expertise in satisfactorily producing the class 41 transmitter—would be met. Honeywell agreed to keep its representatives at the factory in Poland and to provide training in the United States for Polish workers until the quality tests were acceptable. Additional training or representations over and above the specified limit to achieve good acceptance units would be at Honeywell's expense.

Honeywell guaranteed that the class 41 transmitter could be manufactured through the use of its know-how. Proof of the transfer would be the successful production of 75 transmitters. If necessary to meet this end, Honeywell employee services would be available to remain in Poland beyond the scheduled 60 calendar days in 1976, at Honeywell's expense. If Honeywell came to feel that it would not be reasonably possible to manufacture these items in the course of the agreement, the company had the option of informing Metronex of its conclusion through a written notice. Upon receipt of such a notice, Metronex could either accept the Honeywell transfer as satisfactory or accept the cancellation and send both the special testing equipment and the transmitter acceptance kits back to Honeywell after Honeywell reimbursed Metronex for all payments to that date.

Host. Metronex's control element, on the other hand, involved the terms of the agreement itself. Honeywell agreed to transfer certain technology to Poland and to work with the Poles until that technology had been satisfactorily transferred. This required that the information being transferred be correct and suitable, and that the training sessions be conducted in such a way that the skills involved in this manufacturing process were passed along. Because Poland would be purchasing detail parts as well as the sensor units for the duration of the contract, the profit aspect provided another form of control over Honeywell.

It is interesting to note here, however, that the leverage Honeywell possessed through its proprietary claim to the sensor and class 41 technology was not sufficient to prevent Poland from demanding, and receiving, the updated version of this technology at the end of the contract. Honeywell would probably have much preferred to transfer an earlier version of the technology. However, the prospective sales to Eastern Europe apparently were enough of an enticement to induce Honeywell to proceed with the agreement despite its reluctance to part with its latest technology in this area.

Metronex guaranteed not to utilize the know-how involved in the class 41 transmitter, including the sensor assemblies, for any other purpose than the manufacture of these transmitters in Poland. It also agreed not to sell or otherwise dispose of the transmitters outside the specified countries within the Eastern bloc during the period of the agreement or to service such transmitters outside that area without written approval from Honeywell. Metronex also agreed to inform Honeywell of any improvements made regarding these transmitters. Poland had to meet certain responsibilities before progress could be initiated in other areas. For example, Honeywell insisted that equipment installation be completed before the next level of training commenced.

Inspection Privileges

Domestic Company. Inspections were to be utilized by both parties. Poland was to provide Honeywell with two or three samples per year and monthly quality reports on the factory. At least one representative from Honeywell would visit the installation in Poland annually. Inspections pertained to the product being manufactured or assembled in Poland, and Honeywell would inspect five samples annually during the course of the agreement.

Host. Inspections were also to be done on the items shipped to Poland by Honeywell. If any flaws or defects were found in these items, Honeywell was either to replace them or refund the purchase price and Poland was to return the items, with no penalties involved on either side.

Patent/Trademark Defense

Patent and trademark protections involving the class 41 transmitter had several dimensions. First, Honeywell restricted the use of the sensor and all other items involved with the class 41 transmitter to only this product. Metronex was permitted to manufacture, utilize, and sell only this transmitter and was not to use this technology in other forms. Second, should the issue arise, Honeywell agreed to provide defense information to Metronex to defend the patent if a third party contested it.

The only trademark Honeywell provided to Poland under this agreement was the following statement:

> Made in Poland
> MERA-Pnefal under license from
> Honeywell, Inc., U.S.A.

Poland could not use any other indication of Honeywell's name or involvement.

Legal Provisions

External protection for both parties included stipulations in the agreement that arbitration proceedings be based in London, Great Britain, if they should become necessary. Such arbitration would be held in the London Court of Arbitration, using the English language.

The agreement also included a force majeure clause but had no provision for court involvement beyond that of patent defense or arbitration.

Project Assessment

Factors Expected to Make the Project Easier than Usual

The proprietary nature of the technology involved in this transfer required that Honeywell and Poland work together, especially since Poland would be unable to obtain this technology from any other source. Furthermore, since high-technology fields are often rapidly outdated, this opportunity provided Honeywell with the means to utilize its technology. The transfer, while involving intricate items, did concentrate on small, rather easily managed equipment packages—unlike some transfers that involve massive amounts or sizes of equipment. The high skill level required to manufacture the transmitters resulted in highly trained and skillful employees on both sides of the ocean. Furthermore, the relationship with Poland had been paved by numerous agreements between the Poles and the European subsidiaries of Honeywell; by the time the U.S. company was involved, both groups were relatively familiar with each other.

Factors Expected to Make the Project More Difficult than Usual

The major concern in this area would probably be the high skill and precise techniques that were required to manufacture the class 41 transmitters to exact specifications.

Expectations for the Future

Honeywell felt that this project could provide access to the Eastern European market, and thus saw the project as a stepping stone to other agreements, as well as an important agreement in its own right.

Westinghouse Electric Corporation

Description

Westinghouse Electric Corporation agreed to design, equip, and put into operation a plant for power semiconductors in Eastern Europe through an agreement with the Polish People's Republic. The $9.7 million contract, signed on March 12, 1973, and effective on April 1, 1974, was arranged through Westinghouse's Semiconductor Division in Youngwood, Pennsylvania. This was the largest sale of technology ever made by Westinghouse to that date, and it was the first agreement of its kind between a U.S. power semiconductor manufacturer and an Eastern European country. In the contract, Westinghouse agreed to put into operation a manufacturing plant capable of producing a minimum of 200,000 rectifiers and thyristors annually on a two-shift basis. The production facility in Poland would be located in the Lamina complex, an industrial area 10 to 20 miles south of Warsaw in the town of Piaseczno. The agreement between Westinghouse and Poland would last for five years, with production at the facility expected to be under way by 1976, or 32 months after the signing of the contract.

Background

Domestic Company. The greatest share of corporate earnings in Westinghouse comes from the sale of products, not technology per se. However, when this avenue is not particularly feasible, such as when a lack of hard currency negates product sales to any appreciable extent, or when a country is more concerned with using its capital to develop such products itself, Westinghouse agrees to sell its technology.[34]

The corporate structure itself reflects this willingness to sell production know-how in the existence of an international licensing division—a separate profit center whose sale of licensing includes production technology as well as patent rights and documentation. Under this arrangement, the licensing division, together with the manufacturing division, negotiates all international contracts, but the individual divisions (the Semiconductor Division, in this case) are then responsible for carrying out the details of the contract. Consistency is achieved by having personnel from that respective division on the negotiating team when the contract is being drawn up.

Westinghouse does realize the risks incurred by a firm that establishes potentially viable competitors through technology transfers. However, corporate officials in the Semiconductor Division at Westinghouse have pointed out that the power semiconductor industry requires a high capital outlay in equipment and research and development, but the market for the final product is relatively low and competition within that market stiff. Thus, profits in such an industry depend either upon pricing the final product high enough to recoup the development costs or on selling the technology itself. If the product cannot be sold, Westinghouse sometimes agrees to sell the technology. The licensing center within Westinghouse arranges suitable and mutually advantageous technology transfer sales.

Westinghouse feels that its approach to technology transfers such as that involved in Poland does not necessarily create competitors. Howard Wolff interviewed former Semiconductor Division General Manager Maurice G. Sardi who noted that

> "The benefits of selling technology are sufficient to offset the disadvantage of setting up competitors." The way Sardi explains it, "When negotiating the contract, it was necessary to freeze the technology at a particular point four years ago. But since then, Westinghouse has been continuing its own R&D; we are now four years smarter. I can sell technology through, let's say, phase one. But it would take the same resources to get to phase two as it took to get from zero to one, and there are no option clauses in the contract under which we would provide the next phase. Moreover, the purchasers want to advance themselves. They now have the core, but how they perpetuate it is their problem."
> The division's operations manager, Mickey Turner, adds, "It takes a long time to learn the semiconductor business, and it will take many years to get good yields. We just showed them how."[35]

Westinghouse further protects its technology by limiting its transfer contracts to five years duration. This way, the technology is provided at that stage one, but the updating period is relatively short.

Another aspect of Westinghouse's approach to technological transfers is that, although the corporation will not always extend trademark rights to its partner, it virtually guarantees that the transfer will be successful. Westinghouse expresses confidence that its partner will be able to produce the desired product meeting certain specifications with at least the minimum capacity described in the agreement. Any situation wherein Westinghouse is not able to adequately transfer the technology as agreed would reflect adversely upon the corporation's ability as a seller of technology and might jeopardize future profits from this type of transfer.

The standard corporate approach to such transfers may include a buy-back arrangement. This is logical when the trademark has been provided, but it is an option whenever the trademark has not been part of the transfer package. In general, reluctance to agree to a buy-back arrangement usually stems from the impetus that prompted the technology sale in the first place. On the other hand, if the product can be repurchased by Westinghouse at a desirable price at a point when Westinghouse needs an additional product source, the buy-back arrangement is usually present. Even when such a need is uncertain, Westinghouse often includes an optional buy-back possibility.

Westinghouse's customary policy regarding patents is to include all necessary patents with the technology transfer for the period of the agreement, which is usually five years. However, it is possible under mutual agreement to extend the patent license for additional periods of time as convenient. At the conclusion of the stated time period, however, all rights revert back to Westinghouse, although the purchaser maintains full rights to the production of the product.

Despite the magnitude of Westinghouse's international contacts, technology transfers account for only a small percentage of the corporation's overall earnings. It has been stated that Westinghouse has had more than 200 licensing arrangements in more than 40 countries wherein the licensing involvement included technology transfer, patent rights, and production know-how. Of the corporation's $6 billion sales, $35 million (less than 1 percent) came from licensing.[36] However, in terms of profits, Hayden noted that the transfer of technology is virtually pure profit. The overhead involved in most transfers is relatively low because the technology has already been developed; thus, more than one-fifth of the corporation's net income comes from licensing fees and royalties.[37] This is not just a desirable luxury, however, but a means of existence.

Some technologies—including semiconductor technology—change so often that sellers must be able to profit quickly from their latest innovations. Because such changes tend to negate patent rights, innovations spread quickly throughout the entire industry. Thus, unless the technology is sold quickly, it is likely to become obsolete and generally worthless to its creating company.

Westinghouse, based on past experience, expressed confidence in its ability to satisfactorily effect a technology transfer anywhere in the world. However, the project in Poland was the largest transfer of semiconductor information ever undertaken at that time by Westinghouse, and the magnitude alone carried a high degree of responsibility. Prior to the signing of the Polish agreement regarding power semiconductors, the Westinghouse Semiconductor Division was involved in two similar transfers in Puerto Rico and Le Mans, France. In fact, it was hoped that the assembly plant in France would be a source of materials or subassemblies for the Polish plant until production there was totally complete. The Polish contract marked the first time that Westinghouse would transfer the processing and highly technical aspects of this technology, thus constituting a new challenge for the corporation.

Westinghouse transferred the information necessary for Poland to produce thyristors and rectifiers. The trade name for thyristors is silicon-controlled rectifiers (SCRs). These solid-state semiconductors are used to convert alternating current to direct current in industrial applications ranging from welders, motor controls, power supplies, and electroplating equipment to the large equipment used in mills and mining. The thyristors to be produced at the new facility would operate from 40A to 1500A, a fairly extensive range. The distinction between a rectifier (power variety) and a silicon-controlled rectifier is that a basic rectifier converts power, while a silicon-controlled rectifier not only converts power but controls it.

Westinghouse was not the only company with such a product. There were about 12 manufacturers of semiconductors around the world, but Westinghouse and General Electric had the major market share of power semiconductors. Though, Westinghouse was not the only source of this product, it had developed more efficient systems for delivering the product.

These systems were then in use in both the stateside Youngwood location and in the Le Mans, France, installation. The system guaranteed delivery of high-power semiconductor devices in as little as 45 to 54 hours, contrasted with the four to six weeks delivery times quoted by other firms. This new system also made it possible to "bank" partially complete semiconductors and to fill orders from this bank. Thus, semiconductors from the Youngwood facility could be used to help start up the Polish facility. The partially complete semiconductor devices would help the mastery of each new manufacturing step.

As mentioned earlier, unlike some technological areas where innovation is relatively slow, the semiconductor field appears to be in a state of constant revision and improvement, and technology that is only six months old may already be obsolete. It has been noted that the success of a semiconductor manufacturer depends almost entirely on its technological innovation ability;[38] a firm using the latest production technology may be able to cut costs by as much as 40 percent. Thus, Westinghouse's new production techniques provided an added value to its semiconductors over others.

The manufacture of the silicon rectifiers and thyristors created a number of areas wherein improved production techniques added to the efficiency of the process. Silicon, used because of its ability to withstand high temperatures, is a poor conductor of electricity and must be batched or blended with other ingredients, such as boron or phosphorous. The outcome of this process is a silicon wafer, which must then be incorporated into the final packaged product. All these steps in production, however, provide a number of areas where efficient manufacturing methods greatly help to reduce costs of production.[39]

Host. Westinghouse knew that the manufacture of semiconductors was not new to Poland. But the Polish facility was 20 to 25 years old and had been used to manufacture low-power semiconductors. Westinghouse felt that Unitra, the Polish manufacturing facility, was turning out a product that was not suitable for many uses. Furthermore, Westinghouse felt the production methods of the Poles were highly inefficient in comparison to modern procedures.

The agreement with Poland fell under the Polish Ministry of the Machine Industry, which also oversees the electronics segment of the Polish industry, the Unitra Electronics Industry Union. This is the largest single electronics producer in Poland with a 1975 volume of $1.1 billion and an annual rate of growth that averages between 15 and 20 percent. The Poles had sought a number of foreign licenses, of which the Westinghouse contract was only one example. In semiconductor products, 65 percent of Unitra's total production was based on foreign licenses, although overall production involved only 28 percent of foreign input. The five-year plan covering the years 1971-75 included an objective to double the output of the electronics sector of the Polish industry.[40]

Transfer of Technology

Turn-key. The building that would house the manufacturing was already standing, but Westinghouse was responsible for suggesting modifications to the actual building shell, which would have a 30,000-square-foot manufacturing area when completed. Modifications included air conditioning, lighting, and so forth—even suggesting the proper finish for the floors. Westinghouse was to specify, supply, and oversee the installation of all production equipment. Much of this equipment would be purchased by Westinghouse from other suppliers. Some of the equipment included diffusion furnaces, silicon slicers, lapping machines, photochemical equipment, presses, and welders as well as the semiconductor testing equipment. Initially, the plant in Poland would meet its silicon requirements from Western sources, eventually manufacturing its own. No arrangement was made to purchase complete semiconductors from either the Youngwood or the Le Mans facility prior to the start-up of production in Poland; it was assumed that the Poles would use Western European semiconductor producers to meet their immediate requirements. Westinghouse did note, however, that the corporation would be able to help meet any shortages that might arise upon request.

Information and Know-how. Westinghouse provided Unitra with its 1973 state-of-the-art technology for power semiconductor manufacturing—the latest technology that it had in that area at that time. It also agreed to provide Unitra with access to any commercial modifications regarding either technical data or production methods made by Westinghouse during that five-year period, within 60 days from the time that Westinghouse put such modifications into current production. This provision included all technology that Westinghouse might acquire and was reciprocal in that Westinghouse was to be informed of any revisions that Poland might make in the product.

Support services that accompanied the manufacturing information included quality control, incoming inspection, ratings and reliability, calibration, stores, production control, inventory control, base brazing, air-/water-cooled assemblies, mask duplicating, and lead and wire feeder operation.

The approach used by Westinghouse's semiconductor division to effect this technology transfer was extremely detailed, with each step scrutinized carefully and the process figuratively frozen at each point. The semiconductor manufacturing process had been fully categorized by function and by material and equipment needed. This characterization was in the form of a computer program that transferred each manufacturing operation on a step-by-step basis from the Youngwood facility in Pennsylvania to the Lamina factory in Poland. The person in charge of this massive operation was Donald A. Walczak, manager of project engineering for the Westinghouse Semiconductor Division. Westinghouse established a time frame of 32 months from the start of the project until production was operating on its own. To accomplish all tasks within that time frame, the work was divided into three phases: first, preparation—modification of the building to appropriate standards; second, implementation—purchase and shipping of equipment and commencement of training; third, finalization—completion of equipment installation and the running of acceptance tests. The approach was extremely methodological. All equipment was consolidated at the Youngwood facility, and acceptance tests were even run on the equipment there in the materials development laboratory. After these tests, economical containerized sea shipments were sent from Pittsburgh to Poland according to a predetermined timetable, so that the equipment would arrive only when the facility in Poland was ready for it. This approach was practically the ultimate in inventory control. It provided for the complete equipping of each work station including all necessary tools, spare parts, and even service requirements.

The installation at the factory was done sequentially so that each step in the process was completed in chronological order—that is, following the manufacturing steps required to produce the finished devices. Each item was listed alphabetically by function, right down to the number of wastebaskets to be supplied.

According to the timetable, training would commence 15 months after the start of the agreement and would be considered complete only when all acceptance tests were concluded. Additional training was possible, however, especially when any updating was required. Even the project's conclusion was detailed as to departure schedule and the "weaning" of the managers so that the facility was left on its own in a systematic fashion. The contract also included details as to the qualifications required of all Polish specialists to be trained.

Management. No specific general managerial responsibilities were listed in the agreement, although some were indirectly included under the training descriptions laid out in the contract.

Patents. The Westinghouse agreement actually contained two contracts. The first was the Technical Assistance Agreement, which laid out all the procedures and training schedules as well as the product documentation to be transferred. The second contract was the Patent License Agreement, which included all stipulations accompanying the use of the patents involved in the manufacture of the power semiconductors.

Westinghouse granted Poland a nonexclusive license to manufacture specific thyristors and rectifiers in Poland and to sell those products in Poland and the surrounding Eastern European nations during the course of the five-year agreement. In fact, Westinghouse granted Unitra all patents for worldwide sales except those for the United States and Canada. In a general sense, Unitra would be able to market its products in those two areas, too; but, by not providing the rights to the patents for those areas, Westinghouse technically reserved the right to bring a patent infringement suit against Unitra if Poland attempted to sell its products in those countries. The patents revert to Westinghouse at the conclusion of the agreement, although the five-year deadline could be extended for another three years. After the termination of the agreement, Poland may continue to use the patent information without charge.

Trademarks. Westinghouse did not allow Poland the use of its trademark in this agreement, although Poland was granted the right to use the phrase "licensed by Westinghouse" for the duration of the agreement, but not after that point. Any indication of Westinghouse's involvement in particular products must be approved by the corporation in writing. At the conclusion of the agreement, Poland must cease to note any involvement by Westinghouse, although the extension of the patent agreement would allow Poland to advertise the alliance for an additional three years beyond the conclusion of the offer portion of the agreement.

Buy-back. Westinghouse did not include any buy-back arrangements per se in this agreement. However, after the agreement was signed, an arrangement was made whereby Westinghouse indicated a willingness to purchase some items if the product's quality, timeliness, and price were satisfactory to all parties concerned. Westinghouse also retained the option of buying back some parts as well as complete products.

Responsibilities

Domestic Company. The contracts detailed an extensive list of Westinghouse's responsibilities regarding this project. These were contractually summarized as "Westinghouse will specify, supply, and oversee installation of all production equipment, train plant supervisors and assist in the start-up of the plant." Interestingly, the contract also included a brief description of what Westinghouse was not responsible for.

Host. The Technical Assistance Agreement clearly stated all of the Unitra's responsibilities as purchaser. These are as follows:

> Generally, to provide all labor, materials, supplies, information and services for construction and operation of the Plant and all activities related to it which Westinghouse is not obligated to provide by this Agreement, including, without limits:
>
> 1. specification of electrical power, gas, water, sewers, or other existing or readily available services or utilities;
> 2. electrical characters, general statement regarding the reliability of supply and frequency of outages, if any;
> 3. electrical characteristics for power and lighting within the Plant;
> 4. architectural-engineering design and construction of the plant, sources of service, points of entry of service, using metric measures, in the English language, drawings and specifications to be forwarded to Westinghouse for comment;

5. compliance with regulations of local or other government agencies as to safety, employment, hours of work or other matters;
6. facilities for medical attention;
7. security or fire protection;
8. office furniture, etc.;
9. tables, benches, etc.;
10. communications systems;
11. lighting;
12. external transportation facilities;
13. installation of manufacturing or other required equipment;
14. modifications of plant necessary to make plant compatible with specifications;
15. labor necessary to operate plant;
16. administrative personnel;
17. technical and engineering staff to install equipment, tooling and operate plant;

in a proper and timely manner.

Corporate Perception of Benefits to be Gained from Agreement

Westinghouse's desire for the Polish contract is a reflection of the firm's overall market strategy. As noted earlier, when it is possible or feasible to do so, the corporation will sell its products; however, if the sale of those products is unlikely, then Westinghouse may opt to sell its technology.[41]

In this situation, Westinghouse apparently favored the transfer of its technology for several specific reasons. First, the hard currency shortage in Eastern Europe had been a barrier to the sale of Westinghouse products there, so it appeared that the sale of technology was the only sale of any extent that could take place. In addition, there was a tendency on the part of the Eastern European bloc to purchase technology from the United States. Though products imported from Western European facilities were more accessible to Eastern Europe, they were more expensive than those that could be produced locally. Thus, while the doors to product sales were generally closed to Westinghouse, the doors to technology purchases were invitingly open.

Consequently, Westinghouse sought to obtain income from the fees generated by this technology transfer. The corporation stated that these fees were viewed as profit, but their basis stemmed from the need to take advantage of the product before obsolescence set in, as well as the need to compensate for the high development costs accompanying such products. The fees, in this case, were also partly equivalent to royalties since no product auditing was to be done.

The second major reason for Westinghouse's desire for a Polish agreement lay in the possibility that additional doors would open for the company both in Poland and throughout Eastern Europe. The most obvious outgrowth would be additional semiconductor facilities, perhaps different but related to the products produced at Lamina. However, Westinghouse also saw an opportunity to diversify its technological range throughout Eastern Europe, perhaps expanding to marine equipment and air conditioning.

A third reason supporting the Unitra agreement involved the establishment of a possible source for Westinghouse for either completed products or raw materials. The agreement also created a potential for sales of semiconductors to the Polish facility, especially prior to the onset of Lamina's own production—though Westinghouse did feel that the possibility of such purchases was unlikely beyond the purchase of some partially completed semiconductors from the bank in Le Mans. These would serve more as training aids rather than be used to actually fill product needs.

Negotiations

Parties Involved

The Westinghouse executives involved in the negotiations with Poland were headed by Paul E. Lego, at that time general manager of the Semiconductor Division. The negotiating team included Donald A. Walczak, project manager of the transfer operation of the Overseas Licensing group; Douglas D. Danforth, executive vice-president of components and materials; and C. E. K. Fox, manager of the manufacturing development laboratory.

Westinghouse negotiated with Unitra, the foreign trade arm of the Polish Electronic Industry Association. Polish principals included Jerzy Milobedzki, director of Unitra; Aleksander Lobodziec, general director of Lamina; Darious Piwowarczyk, technical director, Lamina; Jerzy Szyjko, manager, semiconductor operations for Lamina; and Jerzy Michalowicz, manager of research and development for Lamina.

Initiating Contact

Unitra made the first inquiry to Westinghouse by letter in the late 1960s. Poland was interested in Westinghouse because of the corporation's strong position in power semiconductor technology.

Jerzy Bilip, Unitra's technical director, stated that Westinghouse was selected as a partner because of its "very strong know-how."[42] The main competition for the contract was from the Siemens Corporation of West Germany, but Westinghouse was preferred because of its lead in manufacturing development and product quality.

Vice-president Doug Danforth noted that, at that time, Westinghouse held 500 of the 2,000 patents generally considered to form the basis of all semiconductor manufacturing technology.[43] Poland was also interested in Westinghouse because of the corporation's experience in international technology transfer; it was felt that this experience would enable the company to successfully accomplish the transfer. However, at that time, a permit was needed from the Department of Commerce and this permission was difficult to obtain. By November 1972 when the proposal was reinstated, Poland had been awarded most-favored-nation status and had received Eximbank credits. Business International noted that "the fall-out from the current U.S.-Soviet detente has made it easier for U.S. companies with critical technology to obtain validated export licenses from the Commerce Department."[44] It also appeared that the Polish commercial counselor's office in New York was instrumental in helping Westinghouse obtain the necessary export permission from the Office of Export Administration arm of the Commerce Department.

Length of Negotiations

Almost two years passed after negotiations were reinstated and before the contracts were signed on March 12, 1973. Since U.S. government approval was not a concern at that time, those two years provided opportunities for ironing out some of the contractual differences between the Polish representatives and the Westinghouse team, as well as a chance for both sides to become acquainted with one another. Both the Technical Assistance Agreement and a Patent Licensing Agreement were officially written in English. Table 3.1 outlines events transpiring approximately two years prior to the agreement's effective date.

Issues and Obstacles

Equipment Costs. Westinghouse wanted to separate equipment costs from other contractual payments, thus preserving the corporation's anticipated fees from any fluctuations in equipment purchases or availability. However, Unitra insisted that the agreement include equipment purchases at set prices. The Poles felt that such purchases were then the responsibility of the U.S. corporation. If the firm spent more than the prescribed amount, it was their concern. However, if the purchases amounted to less than that amount, any

TABLE 3.1

Sequence of Major Events of the Westinghouse-Unitra Project

Date	Event
January 1972	SCD decision to go
April 1972	Meeting with U.S. Department of Commerce
May 1972	First validated license approved
May 1972	Official inquiry received from Poland
June 1972	Letter of intent signed
August 1972	Proposal submitted to Poland
October 1972	Invitation received from Poland
November 1972	SCD trip to Warsaw
December 1972	Unitra technical personnel visit SCD
January 1973	Unitra agreement finalized
February 1973	Final negotiations started
March 12, 1973	Unitra agreement signed in Warsaw
June 1973	U.S. approval of equipment license
September 15, 1973	Semiconductor samples shipped to Unitra
September 21, 1973	COCOM approval of equipment license
September 27, 1973	Notification of government approvals delivered to Unitra
January 30, 1974	Loan agreement signed between Exim Chemical Bank and Bank Handlowy
April 1, 1974	Effective date of agreement

remaining funds were to be returned to Poland. This issue had a bearing on the final format of the agreement involving addenda A and B, which included equipment costs. These were added during the final stages of the negotiations.

Prices quoted in the agreement were general estimates based on actual prices (as far as possible) plus an inflation factor. Pricing out the project took from January to October 1972. Westinghouse cautions any firms involved in such negotiations to know equipment prices or be able to accurately estimate those expenses.

Price negotiations were detailed and difficult. One negotiating tactic used by Unitra was to ask the price of an item and then consider the agreement without that item (and the accompanying price). Then the item (and price) were put back, but sometimes different amounts resulted in confusion, with Unitra representatives watchful for the best advantage.

Buy-back Arrangements. Westinghouse was not desirous of including a buy-back arrangement in this contract and apparently had enough leverage during the negotiations to avoid a buy-back clause as one of the contract terms. Presumably, this was due to the favorable reputation possessed by the corporation in the area of silicon-controlled rectifiers. However, a flexible option was drawn up as an advantageous option to both parties, but it was not included as part of the official agreement. The option stated that product purchases could be made by Westinghouse should the product quality, timing, and price be agreeable to all parties involved.

Special Development. Westinghouse agreed to revise one of its basic products to the extent that it was most suitable for Poland's uses, even if the corporation could not use the product elsewhere. This was feasible because of the lack of a required buy-back clause—that is, Westinghouse could hardly have been expected to agree to such an adaption if the corporation had to repurchase those semiconductors. At the same time, the revision did not totally rule out the use of the Unitra facility as a future source of semiconductors for Westinghouse, because the revisions pertained only to a portion of the total anticipated production.

Time Frame

Overview

Westinghouse and Unitra signed their agreement on March 12, 1973, but the agreement did not go into effect until April 1, 1974. The period between these two dates was necessary to acquire the appropriate and necessary licenses. Westinghouse required an export license from the U.S. government, while Poland had to obtain import licenses and also enter into a credit agreement with the Eximbank of the United States. The contract became effective 15 days after final approval had been obtained by both parties and written notice to that effect was received by both sides. Hayden notes that a major delay was caused by an incident only indirectly connected with the semiconductor industry.

> While Westinghouse was given almost immediate permission by the U.S. government to sell its SCR technology to Poland, the entire project was delayed a full year by COCOM. The difficulty centered on the United Kingdom's objection to the sale of the furnaces. Apparently the United Kingdom was reciprocating an earlier U.S. objection to the

United Kingdom's export of a similar item to East Europe, an objection that came in the days before the United States liberalized its export controls.[45]

The Technical Assistance Agreement expired on the fifth anniversary of its effective date, or April 1, 1979. No more updating was provided after that date. The Patent Licensing Agreement, however, can be extended for an additional three years and would then expire on April 1, 1982. Under such an extension, Unitra could continue to mark the semiconductors produced in the Lamina facility as "made under license from Westinghouse" until that date. All indication of involvement would cease after that time.

Special Considerations

Phases. The initial portion of the agreement was divided into three phases commencing with the effective date of the agreement and concluding with the acceptance tests. The three phases were as follows:

Phase I:	Planning—April 1974 to October 1974
Phase II:	Implementation—October 1974 to October 1975
Phase III:	Finalization—October 1975 through November 1976

A more detailed description of these phases is given in Table 3.2.

Know-how Transfer. Using an intricate computerized program, Westinghouse divided the technology transfer into small sections. Generally, however, the know-how was to be transferred to Poland under the following schedule: documentation to Poland (April 1, 1974, through December 1974); equipment shipped, installed, and debugged (January 1975 to April 1976); training at Youngwood (May 1975 through November 1975); technical assistance at Lamina (January 1976 through November 1976); and acceptance tests (March 1976 through November 1976).

Payments to Westinghouse. Unitra agreed to make four equal payments to Westinghouse during the course of the first two years of the agreement. The first payment was due April 1, 1974, as the agreement went into effect. The second was to follow in October 1974, the third in October 1975, and the fourth payment was to be made in November 1976. In addition to these payments, letters of credit for the purchase of equipment and supplies were to be filed in January 1974, October 1974, and July 1975. Payment dates generally paralleled stages in the transfer process: the first payment was due at the start of the operation, the second represented the acquisition of

TABLE 3.2

A Guideline of Events Transpiring during the Three Phases of the Westinghouse/Unitra Contract

	Phase I 1974 Preparation						Phase II 1975 Implementation												Phase III 1976 Finalization												
	April	May	June	July	August	September	October	November	December	January	February	March	April	May	June	July	August	September	October	November	December	January	February	March	April	May	June	July	August	September	October
Action	1	2	3	4	5	6	7	8	9	10	11	12	13	14	15	16	17	18	19	20	21	22	23	24	25	26	27	28	29	30	31
Effective date																															
Building requirements	x	x	x																												
Manufacturing layout		x	x	x	x	x	x																								
Equipment and tool list	x	x	x	x	x	x	x	o																							
Drawing specs updated	x	x	x	x	x	x	o																								
Requisitions prepared		x	x	x	x	x	x	o																							
Receive at staging site								x	x	x	x	x	x	x	x																
Set-up belt lines																x	x	x	x												
Ship to Warsaw																		x	x	x											
Installation																			x	x	x	x	x								
Debug equipment																					x	x	x	x	x						
Material spec info												x	x	x																	
Manufacturing info															x	x	x	x	x	x											
Training-site														x	x	x	x	x	x	x											
Manufacturing technical assistance																						x	x	x	x	x	x	x	x	x	x
Acceptance tests run																								x	x	x	x	x	x	x	x
Payments							x												x												
Lamina building complete															x																
Lamina equipment selection		x																													
Letters of credit																x															

o = extension due to rectifier technology

Note: The above schedule does not anticipate delays due to causes beyond Westinghouse's reasonable control—that is, work stoppages, natural disasters, or political actions.

manufacturing information, the third reflected equipment delivery, and the fourth followed the acceptance tests.

Training. The formal training sessions were to commence at Westinghouse's facility in Youngwood, Pennsylvania, no later than 15 months after the onset of the agreement. These training sessions were scheduled to begin in May 1975 and to extend through November 1975, although Westinghouse reserved the right to determine when training was complete and mastered. The total period for this training was not to exceed 50 man-months and was not to be less than 25 man-months.

On-site training in Poland was to accompany the technical assistance provided by Westinghouse; although this was to formally commence in January 1976, Westinghouse personnel would have been in Poland long before that date. This technical assistance training would end with the successful completion of all acceptance tests; it was officially scheduled to be over by November 1976.

Deadlines. The deadlines stated in the agreement were as follows: construction, completed by October 1975 when equipment installation began; production, from January 1976 on; acceptance tests, started by March 1976 and completed 32 months after the start of the agreement, or by November 1976.

Extensions. The only extension formally noted was the three-year optional extension. The Technical Assistance Agreement specified a time period for most interactions. The major time lines are listed below.

Modifications: Westinghouse must notify Unitra of all modifications to the product within 60 days of their introduction into production at the Youngwood plant.

Technical manufacturing assistance: A maximum of 24 man-months of resident Westinghouse management would be supplied; a maximum of 24 man-months of technical experts in various fields would also be supplied by Westinghouse.

List of equipment: Unitra must provide Westinghouse with an alternative list of equipment if the list provided by Westinghouse was not considered desirable within 60 days after the effective date of the agreement.

Supply of equipment: Unitra must establish a letter of credit with a specified bank for the amount of $2.5 million within 30 days after the start of the agreement.

Default: Notification of default must be made within 30 days of such an occurrence; remedy must be made within 60 days or termination results.

Performance warranties: Westinghouse warrants that no later than 32 months from the effective date the plant will produce a mix of usable stud-mounted and Pow-R-Disc products at a rate of 400 pieces per day for five days.

Tests: No later than 32 months from the effective date, a test of the fusion and assembly belt lines will be conducted at the Polish site. The test must demonstrate certain results with specified limits at that time.

Trademarks: Any request to utilize any aspect of the Westinghouse trademark must be made by Unitra to Westinghouse in writing, and the corporation must respond within 90 days to that request.

Termination: In the event of default, termination will result not less than 30 days after the final notice is given.

Financial Flows

Payments

As mentioned previously, the payment arrangement between Westinghouse and Unitra divided the total amount into four payments of equal size that would be transferred to Westinghouse at specified periods, generally coinciding with different stages of the project. The individual payment amounts were not divulged, but if the total package amounted to approximately $9.7 million and equipment purchases were close to $4 million,[46] approximately $6 million would remain to be divided into four payments to Westinghouse. These payments included both the initial disclosure fee for the technology and the estimated royalties.

Westinghouse usually charged a royalty for the duration of its agreement on licensed items produced from its technology. In this case, Westinghouse chose not to abide strictly by its corporate policy of charging a royalty of 3 to 5 percent of the unit price but, instead, arrived at a predetermined lump-sum fee based on a projected annual production capacity of 200,000 devices. This approach removed any need for monthly or annual auditing of the Lamina factory's production by Westinghouse.

Financing for the total project was arranged through several banking operations. The Eximbank financed 45 percent, undisclosed commercial banks in the United States financed 45 percent, and Poland provided the remaining 10 percent in a cash downpayment. (*Business International* reported a U.S. government guarantee on the eight-year Eximbank loan.) The Polish bank involved in the transaction as the obligor was the Handlowy Bank.[47]

Purchases from the Domestic Company

Equipment purchases were listed as though from Westinghouse, but the largest portion was only arranged through Westinghouse with the corporation acting as the prime contractor. These purchases, especially those made during the first six months of the agreement, included diffusion furnaces, silicon slicers, lapping machines, photochemical equipment, presses and welders, as well as semiconductor testing equipment. The partially assembled semiconductors that would be used as training aids and as service or replacement parts, as requested in the agreement, were also included. The total cost of these purchases, nearly $4 million during the course of the project, was instigated with an initial letter of credit in the amount of $2.5 million.

Purchases by the Domestic Company

No major purchases were anticipated by Westinghouse apart from equipment procurement, although some of the expenses of the personnel involved in the project in Poland were the responsibility of the corporation.

Other Financial Flows

A number of contingency arrangements were listed in the Technical Assistance Agreement.

Purchase of replacement equipment: Any damage incurred to the equipment once the responsibility had passed to Unitra although Westinghouse would assist in the repurchase of the equipment. If the purchase amounted to more than $2.5 million, however, another letter of credit must be established first. Westinghouse would also be reimbursed for any expenses incurred in performing this task.

Spare parts: Westinghouse was to recommend a one-year's supply of spare parts for each piece of equipment. Unitra was to have a letter of credit for $300,000, and replacements were to be obtained as needed.

Test completion: The contract contained an arrangement whereby Westinghouse was to perform any acceptance tests if Poland was not capable of completing the tests. In such a case, Westinghouse was to be paid $200,000 for these services. Costs arising from any delays in conducting the tests were to be assumed by Poland, if such delays were beyond Westinghouse's control.

Personnel expenses: Westinghouse was to bear all expenses for travel, moving, and housing for its own personnel and was also to pay for their salaries, living allowances, and other remunerations and expenses. However, Poland was to make available suitable housing and living accommodations and to reimburse Westinghouse personnel for any taxes imposed on them by the Polish government. However, Poland was to absorb all expenses including travel, housing, moving expenses, and living allowances for a maximum of 48 man-months of resident Westinghouse management during the technical assistance portion of the contract. The only expenses incurred by Westinghouse itself during this period with these personnel would be the salaries and other similar remunerations of these employees. Poland was to cover all expenses incurred by its citizens while on training visits to the United States in connection with this agreement.

Personnel

Training Tasks

Domestic Company. The Technical Assistance Contract required that the following personnel be involved with the Unitra project:

Works engineering manager: This engineer works with Unitra to review basic plant construction and design, remains on site throughout the entire period of equipment installation, and as needed thereafter until acceptance tests have been completed.

Assistant works engineering manager: Additional help will be available as needed at the discretion of Westinghouse to assist the engineering manager.

Other personnel: A number of technically qualified personnel will be supplied on a short-term basis by Westinghouse at critical points when workload and activity are high. They may be there during equipment installation. (This does not include equipment technicians paid by the equipment supplier as part of the total cost of the equipment purchase—that is, installation and start-up are conditions of the sale.)

Management services regarding technical manufacturing and manufacturing controls: 24 man-months of resident management will be provided from the period when the equipment is installed until the acceptance tests are completed. An additional 24 man-months of technical experts from various fields to assist in initial adjustment and process specifications will be available.

Westinghouse was to bear all costs in the first three categories related to travel, salaries, and other similar remunerations or expenses. Poland was responsible for providing suitable housing and all necessary work permits as well as paying any Polish tax reimbursements. However, Poland was to pay for all expenses in the fourth category, management services, during the technical assistance period and following equipment installation. Westinghouse would provide salary and related remunerations to individuals involved with management services for a specified time period. However, should these individuals be required to stay in Poland after the acceptance deadline, or more than 32 months after the effective date of the contract, all expenses would be the responsibility of Westinghouse (unless the reason for the delays had been the fault of Unitra and beyond the control of Westinghouse).

Westinghouse did not really know what to expect with regard to staffing and consequently adopted a wait-and-see policy. The company had such a wealth of experience from prior international arrangements that there was a general confidence that no situation would be too difficult to overcome. Some corporate concern over potential language difficulties was reflected by the Polish language lessons that were provided to all Westinghouse personnel directly involved with the project. These individuals studied Polish for an entire year before going to that country.

Host. The Polish personnel to be trained by Westinghouse were graduate engineers or technical supervisors; they included at least one mechanical and one electrical engineer. The total number of trainees was not stated, but no more than six specialists were to be trained at any one time in the Youngwood facility. The total period of training in the United States was to be no less than 25 man-months and no more than 50 man-months. The contract also stated that all specialists must have a command of the English language such that no translation services or interpreters would be required.

Training in the United States had to commence no later than 15 months after the effective date of the agreement. The purpose of the Youngwood training was to acquaint the Polish specialists with the materials, processes, and equipment required for the manufacture of the designated thyristors and rectifiers. No mention was made in the agreement of the training to be provided to the Polish specialists on site in Poland apart from the duties specified for the U.S. personnel who would be teaching.

Poland agreed to pay all expenses incurred by its specialists while being trained in the United States, including salaries, living allowances, traveling expenses, and other remunerations. Any claims or the like that might arise against these individuals would also be covered by Poland. However, the specialists were not responsible for any damages caused while acting under instructions from Westinghouse.

The contract's wording was apparently designed to legally cover a variety of situations that might arise, but it was more of a precautionary measure than a predictive one. The stipulation regarding English language fluency also appeared preventative more than predictive.

Protection

Leverage

Domestic Company. Westinghouse was in an excellent bargaining position throughout the contract's negotiation for several reasons. First, and foremost, the corporation was Unitra's first choice as a source of semiconductor technology. This situation usually exists when the technology is proprietary to one company; however, despite similar technology available elsewhere, Westinghouse was chosen—partly because of the reputation held by U.S. companies in general, and by Westinghouse in particular, for high technology and quality products, and partly because of the higher expertise Westinghouse possessed in efficiently manufacturing silicon-controlled rectifiers. Such process know-how becomes part and parcel of the product itself. Thus, Westinghouse actually did have an element of proprietary control.

Another plus in favor of Westinghouse was the corporation's experience in transferring technology. The corporation could virtually assure the buyer of its ability to transfer manufacturing know-how.

The contract used by Westinghouse provided specific guidelines detailing the responsibilities of the Polish side. This itemization provided leverage to resist Polish pressure to proceed with the next U.S. step while Unitra caught up with its responsibilities—for instance, Westinghouse could refuse to ship certain equipment if the building was not complete.

The entire contract was essentially a statement of the guarantees and responsibilities of Westinghouse, particularly the assurance that the technology would be transferred within a certain time period as illustrated by the successful completion of the acceptance tests.

Host. Poland's leverage was more general and hinged on the fact that this technology transfer was really Westinghouse's only feasible entry into that country, since product purchases were unlikely for the reasons described earlier. Thus, Westinghouse was under an obligation to successfully complete this project if further contracts were desired.

Unitra's immediate leverage over Westinghouse rested on the U.S. company's guarantee that it would transfer to Poland the technology to manufacture a certain number of thyristors and rectifiers annually, meeting certain specifications, within 32 months from the start of the agreement. If this did not occur within that time period, Westinghouse was obliged to stay on in Poland, at the corporation's expense, until such acceptance tests were successfully completed. The payment schedule supported this control because the fee payments coincided with particular stages in the transfer process. Had these stages not been completed on time, the payments would have been withheld by Poland.

Poland's contractual responsibilities provided a framework for Unitra's obligations in the agreement. These were associated with completion of the building and the provision of suitable materials, technical people, and some equipment. Westinghouse was not required to proceed with further stages of the project until Poland had completed its responsibilities for that stage.

Inspection Privileges

The agreement provided for Westinghouse's right of access to the Lamina facility as long as any equipment or materials were still under warranty. This right of access enabled Westinghouse to review Unitra's data on the facility as well as to observe the treatment and maintenance of the equipment.

The acceptance tests might also be thought of as a form of inspection. However, these "inspections" were to cease with the successful completion of the plant. At that point it would be assumed that the manufacturing capability had been adequately transferred and needed no further attention.

Patent/Trademark Defense

The separate Patent Licensing Agreement detailed all rights and privileges accorded to Unitra by Westinghouse regarding all relevant patents. Patents not possessed by Westinghouse had to be purchased separately by Unitra. The nonexclusive license provided to Poland was valid through the five years, plus the additional three years if the extension was granted. However, after this point in time, Poland would have the right to use the information without charge. However, Poland could not pass the information on to a third party, even after the agreement terminated. If this occurred, Poland would have to reimburse Westinghouse accordingly.

Poland was required to protect Westinghouse in any infringement actions. In turn, Westinghouse was to use its best efforts to assist Poland in an infringement situation, but Poland must pay any costs. However, Westinghouse was not required to take any action to assist Poland when that country was involved in an infringement suit as a result of selling a patent to a third party.

The words "licensed by Westinghouse" could be used on the product as long as the agreement was in effect, but not after its completion. Poland was permitted the use of statements regarding its involvement with Westinghouse as long as requests were made to the corporation in writing three months in advance and permission was granted. Since no additional privileges were extended regarding the Westinghouse trademark, no other protection was considered necessary.

Legal Provisions

The Technical Assistance Agreement stated the arbitration procedures for both elements of the contract. Any disputes unable to be settled otherwise would be submitted to the International Chamber of Commerce in Paris, before three arbitrators. The case would be conducted in English, and the verdict was to be handed down within three months of the hearing.

The contract carried a four-part force majeure clause negating liabilities causing delays or disruptions between the parties unless otherwise noted. It covered the following situations: first, acts of God, civil disturbances, accidents, third-party delays (for example, in transportation), labor difficulties; second, governmental actions; third, other delays not the fault of the second party; fourth, delays caused by the second party and measures for dealing with such situations.

Westinghouse apparently took the position that overlapping coverage was better than possibly neglecting some of the contract in one article or another. Thus, several articles emphasized points such as limitation of liability, warranties, and governmental requirements that were, at least in part, covered in other articles.

Project Assessment

Factors Expected to Make the Project Easier than Usual

Westinghouse's past experience in technology transfers provided the basis for the corporation's confidence during this agreement. It guaranteed that the power semiconductor technology included a number of preventative measures, such as lists of guarantees and responsibilities for both parties, the customary provisions for arbitration, force majeure, infringements or third-party indemnity, and procedures involved in possible termination proceedings.

Factors Expected to Make the Project More Difficult than Usual

A technology transfer such as this to a socialist country usually carries several inherent difficulties; this was no exception. Westinghouse was concerned about the security restrictions that might be placed on both parties. A language barrier existed, despite translators. The large scale of the transfer, combined with its intricate technology, could negatively affect the success of the project.

Expectations for the Future

The contract with Poland prompted Westinghouse to consider a number of additional transfers with the Eastern bloc. Corporate officials expressed hope that similar facilities could be built in the People's Republic of China, and they were intent on sounding out the governments of Romania, Bulgaria, and Czechoslovakia. Westinghouse realized that its approach to each of these countries must be in line with that country's overall goals—as Walczak noted, "It takes so long to determine what they want; everything is tied in to five-year plans."[48] Both Romania and Bulgaria already had some semiconductor manufacturing facilities that were directly or indirectly involved with U.S. companies, but they did not possess the type of technology sold to Poland.

ROMANIA

General Tire International Company

Description

The agreement between General Tire International Company and the Romanian State Enterprise of Foreign Trade (ROMCHIM) specified that General Tire (GT) was to design, supervise construction, and put into production a $75 million modern, all-radial steel tire manufacturing plant in Floresti, Romania.

The venture was the first and largest project undertaken by a U.S.-based company in Eastern Europe. The completed tire factory was expected to be the "most modern, efficient tire factory in existence."[49] The agreement, signed on March 1, 1973, stated that the intended production capacity of the plant would be 1 million steel-belted radial truck and passenger tires per year. This agreement preceded the one signed between Romania and Control Data by just a few weeks; unlike Control Data's joint venture, however, General

Tire's involvement was that of a long-term, nonequity, technology transfer contract. General Tire would handle the turn-key responsibility, provide industrial know-how for the initial start-up, and continue to update this tire technology for the ten-year duration of the agreement.

Background

Domestic Company. General Tire's stated approach to its international operations has been one of total flexibility. Its corporate associations have included purely technical assistance contracts, management responsibilities without ownership, minority ownership, and —in a few rare cases—majority ownership situations. In the past, General Tire has had affiliations of one type or another in 35 different countries, with direct ties in 26 nations. Currently, the company has withdrawn from such a variety of contacts, preferring to put more attention into fewer operations so that the involvement in these situations could be more inclusive and all-encompassing. Thus, GT now has subsidiaries in nine countries outside the United States and technical affiliations in an additional ten countries. Selection of the type of involvement depends on the particular situation at hand, GT being flexible to the point where the most appropriate linkage is the one usually adopted. The Romanian tire plant is one example of the extensive involvement included under the technical affiliation category.

General Tire uses the "case by case" sale of technology—that is, licensing only selected products or processes* when the situation or parties request it. Its approach to the transfer of technology is able to be an across-the-board total package involving all tire products manufactured by the firm: passenger, truck, off-the-road, tube or tubeless, bias or radial. For instance, General Tire has been involved in negotiations with Czechoslovakia, a country that wants just the off-the-road portion of GT's tire technology.

General Tire prefers that technical assistance agreements be of a relatively long-term duration (from 10 to 20 years, although this usually does not exceed 10) because the transfer of total tire technology is a continuous project of great magnitude that requires a long period of time, especially with the installation of improvements. Within such a time frame, however, technology is advancing so rapidly that the technology transferred at the start of such an agreement is often revised and updated, if not changed completely, by the end of the

---

*This varies somewhat from the description provided by Hayden who stated that GT's approach is an all-or-nothing philosophy (Eric W. Hayden, Technology Transfer to East Europe: U.S. Corporate Experience [New York: Praeger, 1976], p. 92).

agreement. On the other hand, Romania operates (as do most East European centrally planned economies) on the basis of the state's five-year plans, where little allowance is made for changing technology—a point that became apparent during negotiations (the situation, however, was adjusted to meet the needs of the parties involved).

General Tire's corporate policy will permit its General Tire trademark to be used when its technology is involved without an equity basis. An example of this policy is found in the Romanian agreement.* The company does not plan to repurchase any of the Romanian tires on the basis of its trademark. As corporate officials pointed out, the repurchase or buy-back issue is a consideration based on other factors. However, when the tires being manufactured have passed all required acceptance tests, the buy-back potential is a viable option. Philosophically, GT officials have pointed out that technology transfers are viewed as a sale, similar to product sales, not an investment. As such, the quality of the resulting product is as important as the quality of any respected GT product.

General Tire was established in 1915, long after the other major tire companies. Its market niche was not that of a direct competitor to the others, but as a specialty producer of premium, extra-high-quality tires. To create such a product, General Tire required extensive research and development facilities. However, the cost of such research had to be offset by measures beyond the price of the product, and General Tire chose to sell its resulting technology to continue its research and development. Lacking the desire and the capital to establish 100 percent equity in its overseas ventures, General Tire was the only tire company for 40 years to operate on less than a total equity position, a unique position among tire manufacturers. In fact, it had less than 50 percent equity involvement in many of its operations.

GT's first international contract was the establishment of a small company in Mexico in 1932. Now the company is involved in such diverse areas of the world as India, Tanzania, and Argentina, as well as Romania. General Tire's nonequity experience in international contracts has now placed it in the role of consultant to other companies eager to operate on such a basis. The incentive to establish overseas manufacturing facilities has become one of the chief

---

*Hayden stated that GT did not provide its trademark to the Romanians in this agreement, but GT corporate officers stated that Romania did indeed have the use of the GT trademark as long as the quality control tests were acceptable. (Eric W. Hayden, Technology Transfer to East Europe: U.S. Corporate Experience [New York: Praeger, 1976], p. 95).

means whereby the U.S. tire manufacturers have been able to remain competitive on the international front. Some of the other tire companies, although much larger, had never considered technology transfers, having only operated on a subsidiary or joint venture basis. These companies are now coming into the arena of technology selling, but they are finding it difficult to adjust to the nonequity approach because of their lack of prior experience in this area.

General Tire has transferred to at least 40 different tire or rubber-/plastics-related overseas manufacturing facilities "such documentation as tire dimensions, stock preparation formulas (amount and types of resins, rubber, carbon black, and other additives), and temperatures and duration of vulcanization; plus production techniques and process information relating to compounding, milling, and mixing, calendaring, extruding, beading, actual tire building, curing, final finishing, and quality control."[50]

Host. The tire technology possessed by Romania prior to this agreement was a combination of tire technology from several countries over a period of four decades. In fact, the first tire company to operate in Romania was the U.S.-based Goodrich Tire Company, whose plant was built there in 1933 and was operated by Goodrich until World War II (1939), when it became the fully owned property of the Romanians. Romanian tire technology remained at this level until 1954, when a contract was made with the Dunlop Tire Company to modernize the former Goodrich plant. About ten years later, in 1962, the Italian tire company, Pirelli, built a new plant in Bucharest. Pirelli's approach, however, was strictly that of a turn-key operation with no technology transferred after the initial construction and equipping was completed. Thus, Romania's general output of 2 million tires per year was the combined result of these two tire manufacturing facilities—prior to the construction of the IA (initials that stand for Tire Manufacturing in the Romanian language) Victoria II Factory built at Floresti through General Tire's technology.

General Tire's involvement in Romania demonstrates the company's flexibility in adapting to another company's (or, in this case, country's) terms. The construction of the facility in Romania was necessary because their existing level of tire technology was judged to be "at least twenty years old" according to the firm. General Tire would probably have preferred to export tires to Romania, but this desire was not feasible because of the lack of ready capital in Romania, among other reasons. Building such an extensive tire manufacturing facility appeared justified by reason of the country's projected need for tires, which included 40,000 heavy vehicles to be produced annually by 1975 as well as 50,000 passenger cars. Furthermore, General Tire was informed that Romania intended to use only 60 percent of the tires for domestic use, with the remainder being used for export to East and West European markets. Indeed, General Tire even saw the possibility that some Middle Eastern markets would be lost through the Romanian exports.

As noted by Eric W. Hayden in his book Technology Transfer, the sale of technology by General Tire, including equipment purchases, has provided the company with 15 to 20 percent of its total earnings in recent years.

> In the Romanian project, for coordinating the purchase and shipment of approximately $30 million worth of U.S. equipment, General Tire earned 5 percent, or $1.5 million. (The equipment portion of the project was financed by the U.S. Ex-Im Bank (45 percent, or $13 million), U.S. commercial banks (45 percent), and Romania's trade organization Romchim (10 percent down payment) with which General Tire conducted the commercial negotiations.) In addition, it received an initial disclosure fee of $1 million for its technology, as well as a commitment by the Romanians to pay $500,000 per year for General Tire's "on-going technical service assistance and manufacturing assistance."[51]

General Tire, as a corporate policy, feels that its past involvement with nonequity agreements internationally has helped the company to interact with its partners in such a way that the company has developed "a precious and valuable 'know-how,' the art of listening and getting along with a host of varied attitudes and ideals."[52] Thus, General Tire's proposal to Romania stated the following:

> Because of numerous minority and purely technical association ventures which General Tire has been engaged in during the past forty years, it has had to learn the meaning of harmonious cooperation and become an expert at it. In contrast to companies used to 100 percent ownership with dictatorial attitudes, General offers technical knowhow coupled with human knowhow, which is the most winning combination.[53]

General Tire International Coordinator for the Romanian factory project, Gustav Prekopa, has also noted that "Romania has been for decades, and still is, a European leader in oil production and has vastly developed petrochemical industries for the conversion of this commodity into various end products. Tires are one of these."[54] Romania does have an annual crude oil production of approximately

14 million tons and 20 million cubic meters of natural gas; it is Europe's second largest producer next to the USSR. Its population of 20 million offers both sizable employment potential for such a tire factory and a large domestic market for the output from the tire plant.

Romania's vehicle registration has more than doubled in the past five years. The capacity of the new tire factory is intended to bring the country's existing tire manufacturing facilities up to an annual production figure of 3 million tires.

Prior to the construction of the manufacturing facility involving General Tire, Romania had two other large tire factories, one dating back to the 1930s. These plants were largely responsible for both satisfying internal demand and exports. The tire factory involving General Tire was the first Romanian tire factory to be fully equipped with U.S. machinery since the end of World War II.

Transfer of Technology

Turn-key. To accomplish this technology transfer General Tire was contractually responsible for supplying all turn-key services involved in constructing the plant as far as the blueprints and supervision were concerned. Romania would do the actual construction and building. The company was responsible for equipping the plant, purchasing the bulk of the equipment from the United States, but using locally produced parts if possible. Once in production, Romania was to be responsible for supplying raw materials that met prescribed specifications.

As an example of the magnitude of the project, the length of the factory building was 1700 feet. GT supervisor Oz Perrine, who was responsible for overseeing the construction of the Romanian factory, noted that this factory size superseded many others. For instance, the plant constructed in Tanzania was only 800 feet long. The goal of the Romanians was to be able to produce millions of tires annually, and the factory had been designed with this intent in mind.

Perrine's tasks were to interpret the blueprints and construction data for the project and to ascertain that the building and equipment were proceeding as required. It was the responsibility of the Romanians to install the equipment under the specifications laid down by General Tire.

Information and Know-how. The technical service contract involved two broad aspects. The first encompassed the initial transfer of product technology, including the know-how involved in the actual manufacture of the tires. The second was the ongoing portion, including updated technology and marketing techniques that would commence when the first part was judged completed. These functions can be summarized as follows:

Delivery of the specialized equipment
Design of the factory
Assistance in the supervision of installation
Adjustment and fine tuning of the equipment
Assistance in starting up equipment and lines
Start-up of production
Training of supervisory personnel
Continuous updating of this technology for an initial period of ten years after start-up.[55]

The contract listed these functions and included a bar chart of the anticipated completion times.

The technical service contract, phase two of this transfer, provided for annual services to the tire facility in Romania that extended beyond strictly receiving the updated blueprints or new processes in radial tire manufacture. This contract guaranteed that Romania would receive General Tire expertise in troubleshooting any problems that might occur at any time with the GT technology. It also provided for the training of employees for any new techniques. Since the acceptance tests were to provide an indication that the plant was capable of achieving its objectives regarding tire quality and quantity, any problems in these areas after that point would be Romania's responsibility. Romania paid an annual fee for the services of a number of skilled consultants for a specific number of man-days or hours at specific rates. These services were considered particularly important in the updating of the technology because changes are often reflected in the processing of tires during the manufacturing stages.

Management. No managerial responsibilities were included in this agreement beyond training and construction supervision. However, as in most production setups, some management training was combined with basic production techniques.

Patents. All necessary information was included under this agreement, including all needed patents. However, unlike highly specialized areas, production capacity in tire manufacture is as important as the blueprint type of information, perhaps even more so.

Trademarks. General Tire International made its trademark available in the Romanian project. The contract included the stipulation, however, that the GT logo was not permitted on any tires manufactured in Romania until all acceptance tests had been completed at the factory.

Romania had the right to choose whether the tires would carry this trademark or not. The right to use the General Tire brand name was permitted only for those tires designated for sale in Romania and in limited overseas markets (including all the Eastern bloc countries and several Middle Eastern countries). It was not to be used on tires exported to any other locations. Should Romania choose not to use the General Tire brand name, it was still permitted to use the phrase "made in collaboration with General Tire" on tires prepared for export as well as tires sold within the country or the Eastern bloc. All rights to the use of the General Tire name were to cease at the conclusion of the agreement, after the ten years of the technical service contract had expired.

Buy-back. The Romanians would have preferred adding a buy-back arrangement to the contract. This would have substantially reduced the outflow of hard currency because GT would have taken back some of the manufactured products as payment for services rendered. General Tire has found that buy-back clauses often require the company to put forth additional effort to market those tires. In this instance, GT was able to sidestep this issue during the contract's formation primarily because its chief competitors for the Romanian tire plant were also not interested in having to accept a buy-back arrangement.

Responsibilities

Domestic Company. General Tire's two areas of responsibility were supervision of the plant's construction and provision of certain services as stated in the technical service agreement.

Host. Romania's only contractual responsibilities were to provide the requested payments within the specified time frame.

Corporate Perception of Benefits to be Gained from Agreement

Profit was the initial motivation behind General Tire's involvement in this Romanian project. It was also a venture that complemented GT's prior nonequity experiences and offered an opportunity to apply previous experience on a large scale. Thus, because GT was organized for such endeavors, this project promised a good return on assets currently possessed by the company.

Another prime consideration was that this venture would provide an additional supply of its product without the outflow of equity that would be required if General Tire were to construct a facility on its own. However, because of the lengthy time involved before production would be on-stream, this supply source was viewed as a long-run possibility, not an immediate likelihood.

The radial tire technology sold to Romania was GT's latest and most advanced information. With such recent information, it was possible that Romania might actually become a competitor with General Tire. However, GT's view was that the long-run opportunities in this venture more than compensated for the possible competition that might arise.

General Tire was also interested in penetrating the Eastern Europe market in general, especially through such an enormous undertaking as the Romanian tire factory. General Tire felt that Romania was also in an excellent location to export tires to other areas, with its seaport on the Black Sea and its well-developed network of roads. GT's interest focused particularly on Romania, however, because the corporation felt that the country itself "holds the potential for the future as a sizable and rapidly growing marketplace."[56]

Negotiations

Parties Involved

Domestic Company. Numerous corporate officials from General Tire were involved in the negotiations with Romania. One member of the negotiating team was a Romanian national who had been born in Romania and had lived there until 1965. His knowledge of the language and the culture of Eastern Europe provided a common ground between the two parties.

Host. The negotiating hierarchy in Romania consisted of four levels. At the top was the ministry to which the specific industry belonged. In this instance, the tire industry belonged to the Ministry of the Chemical Industry of Romania. The second level encompassed the State Enterprises for Foreign Trade; these are buyers who act on behalf of those who will later administer the plant. The contacting enterprise interfacing with General Tire was ROMCHIM. The third level was the industrial "centrale," which possessed full technical competence. The centrale involved here was the Tire Rubber Goods and Plastics Administration. These centrales are often aided by design institutes; in this case, the design institute was called IPROCHIM. The fourth and final level was the state organization that would be administering the plant under construction. In this agreement, this was the Victoria, Floresti, tire complex.[57]

Initiating Contact

There a number of ways for a U.S. business or company to initiate negotiations for a manufacturing venture in Eastern Europe. The range extends from seeking entry or introduction to a potential partner through such channels as exhibitions, trade fairs, or mail inquiries, to receiving unsolicited inquiries from foreign trading enterprises already interested in a particular product or company. Romania examined a variety of tire manufacturing companies around the world in its opening quest for a tire plant manufacturer, but Romania's approach was to issue a brief, one-page inquiry to all rubber companies rather than to extend an invitation to just the country's first choice.

GT's past involvement in nonequity situations, combined with the experience and flexibility gained from dealing with a variety of locations and working conditions, led the company to start negotiations with Romania in July 1971. GT was firmly convinced that the company could handle the project.

This led to a deeper interest on the part of both parties, and General Tire conducted several official corporate visits to Romania in late 1971 and early 1972. Its proposal was extended in March 1972 and was one of several bids analyzed by Romanian officials. General Tire's bid was accepted.

Length of Negotiations

The contract between General Tire and Romania took up to 18 months to completely negotiate and was over 300 pages long. It was written in English, although the Romanians held their contractual discussions in Romanian, with translations into English for the benefit of the General Tire negotiators. Considering the details that were laid out, GT did not regard the contract as overly long. In retrospect, GT felt that an even larger document would not have been inappropriate, and 18 months was viewed as a relatively short negotiating period. Prior to the start of the official negotiations, Romanian officials had focused on a number of particularly desirable attributes of the General Tire Company. Timewise, this was beneficial because many of the preliminary investigations had been completed from July 1971 through January 1972, thereby shortening the period of time required to conclude the negotiations. The proposal was extended in March 1972 and signed in March 1973; considering the enormity of the project, this was a surprisingly brief period.

Issues and Obstacles

Contract Format. One of the members of General Tire's negotiating team pointed out that the approach of U.S. companies toward contracts in Eastern Europe often differs from the approach taken by Eastern European countries. This was evident in the emphasis placed on all specific details in the contract by the Eastern European nation. An Eastern European country such as Romania uses the contract as a specific indication of the responsibilities and duties to be carried out. Projects will precisely follow such a document to the letter. Thus, in negotiating, GT officials noted that it was important that the U.S. company pay close attention to contract details, since those details would be considered mandatory throughout the functioning of the agreement. Furthermore, while negotiations usually took place between a company and (in Romania) the State Enterprise for Foreign Trade, implementation involved the buyer or individual factory and the industrial centrale. Hence, contract details had to be clear enough to allow all parties to understand the obligations whether or not they were involved in the initial negotiations. In GT's case, since the negotiations were really held at three levels (the chemical ministry, the state enterprise, and ROMCHIM) all elements of the contract had to be very clearly stated, because implementation would take place on the fourth level, the factory of Floresti.

Competition. The political ramifications of the contract between General Tire and Romania extended beyond the transfer of tire technology. Romania wanted the technology to build steel-belted radials, and General Tire, based in the United States, was experienced in nonequity transfers of tire technology. Among its possible competitors from the United States, General Tire was able to demonstrate past experience in similar agreements as well as a proved ability to be flexible and adapt to a particular situation. However, several other firms based in other countries were also strongly favored for diverse reasons. Several European suppliers had the advantage of knowledge of the country through prior contacts. Pirelli, for example, was a more logical vendor because the company was already operating in Romania and was a familiar name.

Another factor involved in Romania's corporate selection was that country's desire for most-favored-nation (MFN) status from the United States. Courting a U.S. company might have been part of Romania's approach toward being awarded this trading honor. It may not have been a coincidence that the MFN discussions were held about the same time as the General Tire negotiations. Romanian President Nicolae Ceausescu did invite President Nixon to the country for a state visit during which the "magnitude of the General Tire contract" was pointed out.

Another possible reason for the selection of General Tire over other tire companies might have been that General Tire was willing to sell its technology relating to radial tires. Michelin, based in

France, was considered to be the most experienced radial tire company, but this company would not sell its technological process. General Tire was in competition with several other firms possessing generally equivalent technology; if GT had not been willing to supply Romania with its total and latest radial tire technology, another firm would probably have been willing to do so.

Romania was very concerned that it receive the latest technology in radial tire manufacturing as well as continuing updates on that technology. Part of this concern was based on past experiences by East European countries wherein purchases of obsolete equipment or processes had been made that were then rendered less valuable as a result of more recent improvements. Thus, General Tire provided Romania with the information and know-how to manufacture two types of radial tires: the tubeless fabric carcass steel-belted radials used on passenger cars, and the fabric carcass steel-belted and steel carcass belted tire for trucks and other heavier equipment requiring tubes. This transfer package did not include the technology possessed by General Tire for construction of off-the-road tires used by larger vehicles, which are often required in mining or earth-moving operations.

<u>Long-term Commitment</u>. One of the major negotiation hurdles introduced by General Tire was adapting GT's need for a relatively long period of time to the Romanian five-year plan. The solution was to separate the contract into two parts; the first involving the construction and equipping of the factory up to the point of tire production inspection and approval, and the second establishing an ongoing technical service agreement that would run for ten years—but commencing only when the plant's production capability was accepted as being on-stream.

<u>Bias, Price, and Equity Obstacles</u>. In his book, Hayden noted three obstacles that General Tire had to overcome in negotiations with Romania.[58] The first was that of a psychological hangover within the U.S. tire industry dating back to 1964 when the Firestone Tire and Rubber Company was awarded a contract with Romania. However, because the contract had been signed with a Communist country, public outrage—including picketing outside the firm's headquarters in Akron, Ohio—caused this contract to be canceled. To avoid a similar reaction, General Tire sought and received a formal letter from the U.S. government through the Office of International Commercial Relations in the DOC that stated that, as of the Export Expansion Act of 1971, it was government policy to encourage trade between the United States and all nations of the world with which we have diplomatic and trading relations, including most of the Communist countries.

The second obstacle Hayden defined was GT's insistence that the company be paid a technical services fee of $500,000 annually for ten years. "According to General Tire, its turn-key price, exclusive of the annual charge, was 10 percent lower than that of its nearest competitor, a consortium of the United Kingdom's Dunlop and Italy's Pirelli. This brought the total package price to approximately $2 million more than that of its European rivals."[59] Hayden noted that General Tire compromised on two aspects that Romania found objectionable in its European rivals. First, the company agreed to consider revising its nonequity technical assistance agreement into a joint venture involving equity by both partners. Second, General Tire agreed to arrange for a Viennese broker to help Romania dispose of $17 million worth of raw materials to help Romania raise some hard currency with which to pay off the dollar cost of the plant.

The third obstacle was really an offshoot of the second in that the Romanians were quite insistent that General Tire be involved on an equity basis with this manufacturing facility. In contrast, General Tire saw this venture strictly as a nonequity type of technological transfer wherein General Tire would receive fees for its technological and managerial expertise in creating and operating a tire manufacturing facility. Thus, General Tire adopted the policy that such a joint venture might be feasible once the equipment was installed and the technology had initially been transferred. When General Tire learned that its European competitors were also averse to entering into an equity agreement, the company urged that the contract be signed without that issue being resolved. The contract included, however, a joint venture option that would be renegotiable within the first five years of the agreement. GT was concerned during the negotiations that it might be providing free design and equipment advice without being in the final running for the contract. As a result, the company was reluctant to take a stand that would cause negotiations to be terminated on issues that were feasible if not totally desirable.

General Tire's corporate policy on overseas joint ventures had been to own between 25 and 40 percent of the equity because it was felt that, while a higher percentage was too costly, any less denied the company necessary operating leverage. However, corporate policy shifted substantially from this earlier view to a position wherein equity was considered from an investment, rather than a leverage, standpoint. In this case, the company did not want to be involved in an equity joint venture because of investment considerations. Hayden pointed out that the Romanian tire plan was valued at $75 million; thus, 25 percent would equal almost $19 million. General Tire's anticipated revenues from the project at the time of the negotiations were in the area of $8 million: $1.5 million from its equipment commission, $500,000 from supervisory fees, $1 million from the initial technology disclosure fee, and $500,000 annually for ten years (or $5 million) from the servicing and technology update fees.[60] General Tire's equity involvements usually evolved on a smaller scale, such as the $10 million projects that were being negotiated with Somalia and Sudan at that time.

Time Frame

The project commenced on March 1, 1973, and was divided into two portions, with the second part to commence when approvals had been given on the first part. The initial task facing General Tire and Romania was the construction of the factory at Floresti and the equipping of that plant. It was estimated that this would take approximately two years. When the plant was completed the quality acceptance tests would be run; when the tires being produced met those acceptance standards, the first stage would be complete. At this point, the second stage, the Technical Services Contract, would go into effect and run for ten years. Thus, the contract's actual length was ten years plus whatever amount of time was necessary to construct, equip, and complete the acceptance tests at the plant.

Financial Flows

Payments

The total cost of the project for the Romanians was financed through loan agreements with the Eximbank and commercial banks in the United States. Payments to General Tire were in U.S. dollars and were nontaxable by Romania.

Hayden calculated that initial payments to General Tire—excluding equipment purchases but including the commission on those purchases—totaled $2.5 million, or $1.5 million for the equipment commission and an additional $1 million for the initial disclosure and service fee. Annual payments of $0.5 million for the ten-year Technical Service Agreement brought total payments to General Tire by Romania up to approximately $8 million by the conclusion of the contract.[61]

The Romanians agreed to pay a technical service annual fee of $500,000 for updated technological improvements made by GT in its radial tire production over the ten-year period. There was an element of risk here in that Romania could decide that its current level of technology in radial tire production was sufficient and decline to pay the annual fee. Since the agreement had been signed, however, this would be in violation of the contract.

When the company had an equity involvement, GT often requested only an equipment commission or a similar fee for such projects, assuming the technology fees would be transferred via the project's technical service contracts.

This technical service fee might be viewed as a form of license fee, as it could represent a percentage of the sales value of tires produced over the ten-year period. Using a constant figure, as on the GT-Romania project, removed the necessity for GT to audit the tires produced and sold and then compute the percentage accordingly.

Such a predetermined sum is usually preferred by Eastern European countries, where all contract details must be identified. However, such a fee might represent a fairly high payment if production were delayed or slowed. On the other hand, it is conceivable that this fee could be less than an actual sales percentage if the production of the plant were higher than anticipated.

General Tire described its contract with Romania as somewhat standard for the Eastern bloc, including the Soviet Union. It was very precise and was to be followed to the letter. However, there were some nonstandard variations. In most contracts of this type, a percentage of payments (including those initial, up front ones) is often held back until all responsibilities have been completed for that particular stage. If anything was not completed by the specified date in the GT-Romanian agreement, there were stipulations for penalties to be paid back to Romania; however, no portion of the payments was held back in anticipation of this occurrence. On the other hand, the initial technical disclosure fee of $1 million, with the remaining $5 million to be paid over the course of the ten-year period, remained constant. No escalation clause was mentioned in the contract, although inflation or other factors could substantially reduce the actual dollar amount of the payments by the end of the ten-year contract.

Within the Eastern bloc, some countries have more difficulty raising hard currency for foreign trade than others. Romania, however, used its petroleum resources as a base for its capital outlay, since it is the second largest producer of crude oil and natural gas in Europe (the USSR is first). Other countries are not so fortunate. In some instances the methods used to overcome a shortage of hard currency are often the most crucial bargaining points; in some cases, a compensation arrangement is necessary to finance the project. In this instance, GT agreed to help Romania market some products to help with the financing but did not agree to any mandatory product repurchases.

It was anticipated that plant construction would require approximately two years and that the acceptance test would be completed shortly thereafter. Thus, by 1979, General Tire expected to have received complete payment including its commission on all equipment payments.

Purchases from the Domestic Company

Equipment purchases from General Tire were part of the overall agreement and were not considered separately. GT did not anticipate any additional equipment purchases from Romania because it was intended that the equipment supplied would be sufficient for production; all immediate repairs or replacements were covered by manufacturers' warranties. No tire purchases were anticipated, since Romania did have some tire manufacturing facilities within the country already, although it did not possess radial tire technology.

Purchases by the Domestic Company

The agreement stated that Romania would have the responsibility of providing GT expatriates with suitable housing and transportation; thus, the company had no major concern regarding projected expenses in this area.

Personnel

Training Tasks

Domestic Company. The agreement called for a total of 300 man-months to be provided by General Tire. The company's initial expectations were that 12 to 14 people would be used on a permanent basis ("permanent" meaning residing in the country longer than six months but not beyond the completion of the project)—that is, GT employees would be on site in Romania approximately two years.

The two initial fee payments for the equipment and initial technology transfer included provisions for a certain number of man-hours, both in terms of visits particularly relevant to equipment deliveries and adjustments and in terms of longer stays for those assigned to live in Romania for a period of time (for example, during the plant's construction). General Tire was to assume all additional expenses for any man-months exceeding the amount enumerated in the contract. Romania was responsible for providing adequate housing and transportation for all GT employees stationed in Romania.

General Tire anticipated no major problems, since the corporation's experience in adapting to international situations had been extensive, and it was assumed that all obstacles could be overcome or worked out.

Host. A wide range of skills among the Romanian workers was needed because the project ranged from construction of the facility to the operation of the factory. Production training was to be accomplished both in Romania and at GT facilities in the United States, but all expenses were the responsibility of Romania.

General Tire anticipated problems in training some of the Romanian workers because of a lack of appropriate skills. There was also an expectation of fluctuating quality levels. However, GT knew that many of the employees had prior experience in tire production and expected that any problems could be overcome with proper training.

Protection

Leverage

Domestic Company. GT's periods of greatest leverage were during the negotiating stage and during the technical service contract stage. The latter stage would be particularly important if it should happen that certain extremely important improvements were made in radial tire technology. But it was the period between these stages during which GT transferred the greatest body of technology outright. Some industries, particularly those involving intricate electronics, are able to hold back certain portions of technology as insurance that the other party will accomplish all promised responsibilities. However, in the manufacture of radials, most of the information must be passed along immediately. One of the few controls available to General Tire during this construction stage was tied to its receipt of Romania's quarterly payments. Should the payments stop, General Tire's help on the project would also stop.

General Tire International guaranteed to Romania the ability to manufacture radial tires equal to those produced in the United States, assuming that workers at the Romanian plant followed GT's directions and that the raw materials used in production met certain specifications.

General Tire's tire production guarantee extended only through the successful completion of the acceptance tests, which indicated that the Romanians had mastered the initial production skills and the equipment was capable of producing the expected tire quality. On the other hand, the annual technical service fee provided both for the implementing of any improvement that might be developed and for servicing and quality maintenance of equipment.

Host. The payment schedule was a control that Romania could also use to keep General Tire to its commitments. Since the bulk of the payments to General Tire were to be paid while the technical service contract was in effect—that is, after the completion of the plant and while production was in process—the pressure would be on GT to complete the acceptance tests as soon as possible in order to define the end of the contract's first stage and the beginning of the second.

Romania guaranteed, on the other hand, to construct the factory within a certain period of time, to provide employees for both the construction and equipping of the plant, and to supply additional employees with specific abilities or skills to perform the operational functions necessary to move the plant into tire production.

Inspection Privileges

Inspection provisions were established on two levels for the tires produced at the new facility. Tires prepared for export had to meet more rigid qualifications, but all tires had to meet certain specifications.

Patent/Trademark Defense

After the acceptance tests were completed, all tires were permitted to carry the notation that they had been produced in Romania with the help of General Tire. Tires exported to an area where GT had another contract, as stated in the agreement, could not carry the company's official logo; allowing this would make Romania a direct competitor of General Tire in those areas. Those tires manufactured and sold in Romania after the technical service contract was in effect were to carry the General Tire logo if their quality met certain standards. Any tires produced prior to the successful completion of the acceptance tests were not to carry this logo.

Legal Provisions

The agreement provided for several external protection measures beneficial to both parties, such as the inclusion of the force majeure clause and the provision for arbitration should disagreements arise. The arbitration court identified in this agreement was located in Paris; the negotiating language was French.

Project Assessment

Factors Expected to Make the Project Easier than Usual

General Tire's prior experience in nonequity technology transfers was a resource that the company drew upon during negotiations. Romania's desire to have General Tire handle the project and the political environment surrounding this transfer also had salutary effects on the negotiating process.

Factors Expected to Make the Project More Difficult than Usual

The greatest deterrent to this project was its overall size, as it was one of the largest transfers of this type undertaken by General Tire. An accompanying factor was the restrictive, closely controlled nature of the Romanian political sphere, which prevented much direct contact between General Tire and the tire factory.

Expectations for the Future

General Tire was certain that it would be able to transfer this tire technology and felt that the factory in Romania would provide a potential future source of supply, if needed, as well as a central hub from which General Tire's reputation could spread throughout Eastern Europe.

Clark Equipment Company

Clark Equipment Company, headquartered in Buchanan, Michigan, and the Foreign Trade Company Universal Tractor, located in Bucharest, Romania, entered into a series of agreements signed on April 10, 1975, and effective as of October 29, 1975.

Description

Clark Equipment A.G. transferred to Universal Tractor in Romania manufacturing technical assistance, documentation, and patent trademark licenses for power-shift transmissions and torque converters. This technology was purchased from the U.S.-based Clark Equipment Company through a Technical Assistance License and Cooperation Agreement, a Trademark Agreement, and a Marketing Agreement.

The Technical Assistance License and Cooperation Agreement was between Clark Equipment A.G. (CEAG), the European subsidiary of Clark Equipment Company, and Universal Tractor (UT); it provided the transfer mechanism for the information and know-how provided by Clark to Romania. The Trademark Agreement with Universal Tractor was also handled by CEAG. All trademarks and trade names owned by Clark on and in connection with the power shift transmissions and torque converters were to be supplied to Universal Tractor, subject to specified quality controls.

The Marketing Agreement, the third part of the total package between Clark Equipment and Universal Tractor in Romania, contained the same effective date as the other two portions, October 29, 1975, and was to run concurrently with the other agreements. However, this agreement was with Clark International Marketing S.A. (CISMA), the export marketing subsidiary of Clark Equipment Company, which promotes the sale of Clark's products in most of the world markets. Universal Tractor received the services and cooperation of CIMSA in selling and promoting sales of the transmissions and torque converters it produced from Clark technology. CIMSA would also benefit from this arrangement in that Clark would have an additional supply source for these products in Romania, and all promotional endeavors would help to increase Clark Equipment's volume internationally while broadening the awareness of Clark's products. Sales of Clark-related products produced in Romania affected the payments that Clark was to receive from the transfer, as well as the sale of components from Clark to UT for use in the creation of these products. This would be a factor only initially, however, until full production was on-stream. Unlike Poland, Clark would receive no license fee for those transmissions kept in Romania, but it was to receive a fee for all those sold outside that country.

Background

Domestic Company. Clark Equipment describes its worldwide corporate commitment as very high—with hopes of conducting at least 50 percent of its business overseas by 1980. However, it should be noted that the corporation is limited by the amount of equity that it has available and, as well, notes a lack of sophistication that makes it difficult to be innovative. Clark, therefore, uses license agreements as a wedge, which it then drives into a market with the help of its local partner. This type of license agreement includes a know-how package far beyond the kind of information found on blueprints. Clark provides the know-how and manufacturing expertise behind the production of its products because it feels that local production of Clark products will be a must in the future. Eastern Europe, to Clark, represented a large market that was closed if approached via normal marketing channels. Thus, because of Clark's inability to enter such a market, it considered a technology transfer approach. If Clark had been able to enter the Eastern European market in other ways, a technology transfer might not have been tried because of its inherent disadvantages—for example, the creation of competitors and a lack of cost controls. Clark's interest in the Soviet bloc was not limited to the 1960s or 1970s. Clark wanted to enter the Soviet bloc years prior to this agreement with Romania. Immediately after

World War II, the Soviet Union wanted to build heavy axles, and Clark was to provide the turn-key operation to set up such a plant. However, the U.S. government embargoed the shipments of equipment and sold them at auction for less than the letter of credit. After that, Clark adopted a policy of exporting with a German joint venture base. This remained fairly constant until the 1970s when the nonequity agreements were negotiated.

No distinctive features relating to Romania were found in Clark's background that were not already discussed in the Clark/Bumar contract descriptions. The Romanian agreement was signed three years after Clark's effective date with Bumar in Poland and was a sign of marked progress on the part of Clark in penetrating some aspects of the Eastern bloc. Clark's emphasis on axles in Poland and transmissions in Romania was for two reasons. First, Clark was prevented by the Polish agreement from establishing another axle facility within Eastern Europe. Second, the transmissions were a second step toward a more complete Clark manufacturing operation involving total vehicles. Clark envisioned reciprocal trade between Poland and Romania—axles for transmissions and vice versa.

Host. Universal Tractor was described as Romania's foremost manufacturer of tractor equipment. It had facilities for manufacturing power-shift transmissions and torque converters, but it needed the information and technical know-how to be able to produce such items.

Transfer of Technology

Turn-key. Clark was not involved in this project in the true turn-key sense but was expected to provide some of the services often found in connection with turn-key operations. CEAG was to furnish a list of required equipment used by Clark in the production and assembly of the designated products, and to recommend a plant layout. The company would then assist in the placing of men, machines, and materials in the manufacturing plant for the successful production of transmissions and torque converters. Clark also agreed to furnish any special tools at cost to Universal Tractor, at CEAG expense. The facilities were scheduled for completion by 1977 and would cover more than 250,000 square feet.

Information and Know-how. The technical information supplied by Clark was to include all of the following:

Engineering drawings of details and assemblies
Engineering bills of material
Material specifications

Assembly and production test information
Inspection procedures
Process sheets and route sheets
Production, inspection, and assembly tool drawings and fixture drawings
Metallurgical information
Suggested machinery requirements and plant layout
SAE standards and AISI standards and any available cross reference information
Patents
Test programs to establish trademark approval of transmissions and torque converters

It should be noted that Clark provided information on how to manufacture the parts to Clark trademark quality—but not how to design them.

The transmissions supplied to Universal Tractor included model numbers 2000, 3000, 4000, 5000, 8000, and 16000. Mating torque converters included model numbers C170, C5000, C8000, and C16000. The technology supplied by Clark would include all subsequent improvements for subassemblies, component parts, and accessories for use on, or in connection with, the specified transmissions and torque converters together with all spare and renewal parts.

The cost of reproducing the information (two sets) was to be borne by CEAG with one (reproducible) set of drawings. Clark was to furnish all existing engineering changes to UT without further requests or charges to UT covering all modifications or improvements, including all technical information on the product models listed in the agreement. Clark also agreed to answer any UT inquiries and furnish advice relative to the agreement.

CEAG was to supply UT with a list within 120 days of the effective date of the agreement suggesting parts that should initially be manufactured in Romania and those that should be purchased by UT for the first 100 assemblies. Clark agreed to use its best efforts to help UT purchase those components.

Management. As with other arrangements in Eastern Europe, Clark did not plan to take part in the management of the operation apart from skills training in the processes and plant operation.

Patents. The agreement covered not only those patents directly related to the listed products but also all patents that CEAG might acquire during the life of the agreement that related to the particular products whose technology was being transferred to Romania.

Universal Tractor provided CEAG with assurances that Clark Equipment would receive—royalty-free—nonexclusive licenses to any transmissions or torque converters held by UT during the course of the agreement, including access to all inventions, improvements, or patents originating during the time period of the agreement.

Trademarks. Clark U.S.A. authorized Universal Tractor to use its trademarks throughout Eastern Europe, including Poland, Albania, Bulgaria, Czechoslovakia, Romania, the German Democratic Republic, Hungary, Yugoslavia, Mongolia, and the USSR. Before the trademarks could be used, however, all products manufactured in Romania had to qualify under the specified test procedures. In addition to the designated tests, Clark was also permitted access to Universal Tractor premises at all reasonable times, given prior notification, for the purpose of inspecting its products or methods of product manufacturing and quality controls. Clark also maintained control over the usage of the products to ascertain that certain parts were not used for the wrong purposes.

In addition to the Clark insignia, all products manufactured in Romania would carry a designation or indication that the item had been manufactured by Universal Tractor under license from Clark Equipment. This was required and applied to the use of all Clark trademarks or logos in sales or other promotions. Universal Tractor was not allowed to use any other notation that included or was confusingly similar to the Clark trademarks. In addition, Universal Tractor was not to conduct its business in a manner that was harmful to Clark's trademarks or the goodwill that they represented.

Trademarks were provided to Romania on a nonexclusive basis, permitting Clark to use or permit others to use its trademarks concurrently. The fees and financial arrangement concluded between the two parties did not involve any additional payment for the use of the Clark trademarks. The right to the Clark trademarks was for the duration of the agreement, after which time all trademark permission would cease.

Buy-back. Clark had a buy-back clause with Romania, but it was more an option than a necessity and was included to help Romania with its fee payments and to help the country develop its construction industry. As such, the Romanian project was designed to provide mutual advantages for both partners. To help the marketing effort on all sides, mutual cooperation was encouraged; UT and CIMSA agreed to exchange available market forecasts for the maximum sale of their transmissions and converters.

CEAG provided UT with nonexclusive paid-up licenses to make, use, sell, or otherwise dispose of the designated transmissions and torque converters within the prescribed Eastern European area. CIMSA's intent, meanwhile, was to supply all current and future

Clark customers located in Europe, Africa, and Asia with the transmissions and torque converters manufactured in Romania. It was expected that Clark's purchase of these UT products would exceed the payments due from UT to Clark. In fact, it was anticipated at the signing of the agreements that CIMSA's purchases of the model 2420 transmission alone would equal two to three times the total fees paid by UT as stated in the Technical Assistance and Cooperation Agreement.

CEAG agreed not to enter into a similar production agreement with any other country within the defined area for the duration of the agreement. Furthermore, if during this time Clark wished to manufacture any power-shift transmission or torque converter within the Eastern European area not covered by this present agreement with UT in Romania, the matter would be discussed with UT before any manufacturing rights were granted to another manufacturer.

UT was permitted to subcontract the manufacture of subassemblies, parts, or entire models within Romania but only after taking proper precautions regarding product quality maintenance and technology secrecy.

CEAG also granted UT the right—worldwide—to use or sell or otherwise dispose of complete vehicles incorporating the transmissions or torque converters (including replacement parts for the products manufactured under this agreement). UT was given rights to all current and future CEAG patents applicable to the agreement products for the duration of the agreement.

Other Aspects. The opening statement of the Technical Assistant License and Cooperation Agreement emphasized the possibility of a joint development program; it stated "both UT and CEAG are interested in expanding this Agreement over a period of time and will jointly study the feasibility of a joint venture involving the manufacture and sale of the products covered by this Agreement over a period of time and and will jointly study the feasibility of a joint venture involving the manufacture and sale of the products covered by this Agreement and possible other products of the Clark Group of mutual interest." Thus, the possibility of joint product development and/or marketing efforts was anticipated from the beginning of the project, although no specific suggestions or plans were made at the start. One final point, interestingly, was that all three agreements listed the official addresses of both parties. All correspondence involving the agreement—for example, giving or requesting notices—was assured of being sent to the correct address.

Responsibilities

Domestic Company. CEAG's responsibilities included the following:

Providing technology information to UT
Furnishing engineering change information to UT
Furnishing tools to UT at cost
Answering UT inquiries on design and technological information
Indicating initial parts to be manufactured or purchased and supporting those purchases
Arranging for visits of CEAG specialists in Romania
Training UT technical personnel
Furnishing a list of required equipment
Inspecting and testing agreement products
Refraining from entering into similar agreements with others for similar products
CEAG obligation limited to existing information
Allowing UT the right to subcontract
Clark transmissions not included in agreement

CIMSA obligations involved no technical information transfers but did pertain to marketing responsibilities in selling the transmissions and torque converters or buying the Romanian products for Clark's international marketing use.

One of CEAG's first obligations was to provide UT with the initial installment of technical information within 120 days of the effective date of the agreement's acceptance. CEAG would also aid UT in obtaining information or purchasing equivalent components from third-party suppliers for all components and parts required in the manufacture of the designated products, which were purchased by the Clark group from other companies.

Interestingly, the technical information supplied to Romania was to be in French, English, or German. All conversions to the metric system would be done at Romania's expense; however, Clark was to assist UT in specifying equivalent materials to compensate for differences in available resources. Clark had the responsibility for approving all changes or conversions.

Host. Universal Tractor's obligations included the following:

Payment of the initial disclosure fee
Payment of subsequent fees to CEAG (not including sales to the Clark group)
UT was to pay no fee on parts purchased from the Clark group
Calculation of dates and payment of fees

Accountability for sales outside Romania
Availability of UT technicians with appropriate basic skills

To help with CIMSA's responsibilities UT had to demonstrate its willingness to meet all quality, delivery, and price prerequisites during the term of the agreement as initially approved. UT must inform CIMSA of the characteristics of the vehicles for which the transmissions and torque converters were intended, and CIMSA was to advise UT within 30 days of the proper transmissions for those vehicles. CIMSA could not deny advice in this area but was not to be held responsible if UT did not request such advice or if transmission failure was due to improper selection or usage.

Universal Tractor agreed to honor the secrecy of the technical information provided by CEAG not only for the duration of the agreement but also for an additional five years. UT was also to apply its best efforts in manufacturing the maximum number of products commensurate with the markets available to it for those products.

UT agreed to follow the CEAG engineering manual regarding the testing, performance, specifications, and recommendations for manufacturing the transmissions and torque converters covered by the agreement.

Corporate Perception of the Benefits to Be Gained from Agreement

Clark's desire for a Romanian agreement was threefold. First, there was a desire to obtain a manufacturing source for marketing elsewhere, particularly in Eastern Europe. Second, the agreement acted as an entry vehicle for Clark into Eastern Europe. Third, Clark hoped to obtain additional income from UT's payment of the disclosure fee and royalties. It was felt that Romania also wanted this agreement in order to save hard currency, to improve its transmission development, and to obtain more hard currency through exports.

Negotiations

Parties Involved

Clark Equipment's representation, especially those individuals selected by the headquarter's staff to be the company's chief negotiators, met with representatives of both the Romanian government and Universal Tractor.

Initiating Contact

Romania made the initial overture to Clark Equipment, but the contact had been facilitated through Clark's growing relationships in Eastern Europe—stemming from the company's agreement with Bumar in Poland.

Length of Negotiations

The Romanian agreement took two and a half years to conclude.

Issues and Obstacles

Each party was responsible for all permissions and approvals from its own government. CEAG had general approval from the U.S. government prior to the effective date of the contract but had yet to receive formal permission for the license from U.S. and Swiss authorities; UT meanwhile sought approvals from the Romanian government. The effective date of the contract was to be the date on which the final approval was obtained.

CEAG and UT agreed to negotiate on 180- and 140-horsepower transmissions for crawler tractors, and the general agreement was not to become effective before those negotiations were completed. This amendment was attached to the Trademark Agreement between Clark and UT, effective on April 10, 1975, the same date as the agreement itself. In addition to the transmission model series and mating torque converter model series designated in the agreement, this amendment specified that two crawler tractor transmission models (numbers LT7758 and LT7759) were to be added. The amendment also indicated that other countries, upon mutual agreement, could be added in addition to the Eastern European countries designated in the agreement. This would involve only those interested in utilizing the Clark trademark as applied to the products produced in Romania. As with the original statement, UT agreed to furnish representative production samples for trademark tests and qualification.

Time Frame

Overview

The contract ran for a period of ten years from the effective date and could be extended for an additional period if conditions had been agreed upon at least one year prior to the end of the stipulated ten-year-period. However, the ten-year period applied individually to each model produced under the terms of the contract and dated from the start of that model's manufacture in Romania. UT was therefore permitted all rights regarding each model as long as the required fees were paid for the indicated ten-year period.

Special Considerations

The technical agreement was divided into several project phases with recommended time frames. These phases were as follows:

Phase 1: CEAG provides layout; UT sends three engineers to the United States within the first five months
Phase 2: Transmission details will be provided within seven months to UT either within the United States or at an agreed location (where UT is to act under CEAG's advice)
Phase 3: Components not available from the Clark group will be built; Two prototype transmissions will be assembled as a UT activity; CEAG will provide one engineer within 12 months
Phase 4: Lab tests will be conducted using one CEAG prototype; UT will provide one engineer for training within 18 months
Phase 5: Vehicle test will be performed using one prototype transmission from UT; CEAG will provide one engineer within 29 months; Tests will be performed on up to 2,000 hours of work
Phase 6: O-Series preproducts (UT activity) will be produced within 30 months
Phase 7: CEAG will provide final approval certificate, with UT agreement, within the thirtieth month

Financial Flows

Payments

The contract stipulated that all payments and fund transfers would be in U.S. dollars and made to a bank agreed upon in writing, with the exception of the sale of products by UT to CIMSA. CIMSA agreed to sell UT spare, replacement, or renewal parts according to the Clark Central Parts Mulheim Net Original Equipment Manufacturer's Prices, with the same discounts given to other licensees. CIMSA also agreed to sell UT complete torque converters and transmissions, as desired. All payments would be made in full in U.S. dollars via bank transfers through a mutually agreed upon U.S. bank, unless payment was to be in products, which would then be priced with delivery made Free Romanian Border.

Romanian payments to Clark would hopefully be totally in products, not dollars, even though the fees were calculated in dollar amounts. It was anticipated that Clark Equipment would purchase at least an equivalent amount (dollarwise) of transmissions and torque converters during the course of the agreement. However, should Clark not offer to purchase merchandise in an amount at least equivalent to the amount of the disclosure fee, the license/service fees would be waived until purchases totaling the amount of the initial disclosure fee had been made. UT did not have to pay any fees for component parts purchased from Clark that were incorporated into the transmissions or torque converters. In such instances, the cost of those components would be deducted from the established price of the final product.

UT agreed to provide CEAG with a semiannual report of product sales outside Romania; in this case, if Clark's orders had not been equivalent to or above that amount, UT would pay Clark a specified fee in U.S. dollars within 90 days of the end of each reporting period.

Universal Tractor's obligations regarding these fees included paying the disclosure fees to CEAG by the following time periods:

First payment: within 120 days after the effective date of the agreement (after the technical information has been transferred to UT as described under Schedule I, Phase I)

Second payment: within 270 days after the effective date of the agreement (after the technical information has been transferred to UT as described under Schedule I, Phase II)

Four additional payments: within 60 days of each anniversary of the effective date until the total fee has been transferred from UT to CEAG

The license fees owed to CEAG by UT will be calculated as 5 percent of the net price of all sales outside Romania of products manufactured as a result of this agreement (except spare and replacement parts and products sold to Clark Equipment). These fees will be for the rights and licenses granted by CEAG and for its continuing obligations (such as product improvements or technology updates). This payment will be in the form of products and will be calculated on the basis of a mutually negotiated net price. Should the products not be deliverable, UT would then be responsible for paying these fees in U.S. dollars.

Purchases from the Domestic Company

CIMSA agreed to sell UT all required components necessary for the product's manufacture according to the standard CIMSA price list, including discounts used for other licensees.

Purchases by the Domestic Company

The "net CIMSA price" stated in the agreement referred to the price Free Romania Border and was the price at which UT would sell the agreement-related products back to Clark. To achieve its promotional and sales quotas, CIMSA estimated that this price could be no more than 65 percent of the price for the same product as listed in the CIMSA price book at the time of delivery (or, for an equivalent spare replacement or renewal part, 65 percent of the Clark Central Parts Mulheim Net Original Equipment Manufacturer's Prices).

All prices were to be based on the CIMSA Price Book issued by Clark International Marketing S.A. Automotive Division, which lists the prices in effect for Clark transmission and converter assemblies f.o.b. U.S. factory (or ex-factory), as published on January 1 of each year.

Personnel

Training Tasks

Domestic Company. The agreement provided for Clark expertise both in Romania and in training Romanians in the United States. CEAG personnel specialists would be at the disposal of UT for a total of 400 man-days, though no individual visit was to exceed 21 days. Any additional man-days in Romania would be at the expense of UT, including salaries, air passage, local transportation, and hotel accommodations.

Host. An additional 400 man-days of training were provided for UT trainees at Clark manufacturing sites. Under the terms of the agreement, UT would pay for the air passage nearest to and from the place of training, but CEAG would then be responsible for all local transportation and hotel accommodations. UT was to ensure that at least one member of each team would have a good working knowledge of English, but CEAG would also use its best efforts to provide local interpreters.

Protection

Leverage

Domestic Company. To maintain its leverage, Clark built in controls that provided it with some advantages throughout the course of the project. In this project with Romania, Clark's marketing agreement provided a high amount of leverage because both partners were hopeful that CIMSA sales of the transmissions would be high. In such an arrangement, sales would benefit both parties.

One other point of leverage was the fact that, while license fees were accorded to Clark throughout the course of the agreement, the bulk of the payments would be made within the first few years of the agreement, reflecting the transfer of the major portion of the technology. As well, Clark's repurchase intentions, with the award of its trademark and those accompanying controls, helped to ensure ongoing product quality.

Host. Romania's leverage was similar to most other agreements of this type, in that Clark had to live up to its obligations to be awarded its fees. Furthermore, since Clark was hoping to purchase transmissions for its own marketing use, Romania had some control over those purchases. Again, since sales benefited both partners, teamwork was encouraged.

Inspection Privileges

CEAG had the right, but not an obligation, to send one of its representatives to Romania to inspect the products being produced after providing UT with prior notification.

Patent/Trademark Defense

CEAG was responsible for all trademark registration, renewals, and defense; Universal Tractor was responsible for maintenance of its status as a registered user. UT was also responsible for notifying CEAG of any trademark infringements that came to its attention, but proceedings upon such notification were to be at the discretion of CEAG.

Legal Provisions

The arbitration court designated in all three agreements was to be the International Chamber of Commerce in Paris, France, through the common courts of the Arbitration Commission, overseen by three selected arbitrators.

The force majeure clause in all three agreements included any and all unforeseeable, inevitable, or unavoidable events that occurred after the effective date of the agreement, relieving the affected party from liability. Any notification invoking the force majeure had to be made within 15 days of the occurrence of the event (or as soon as possible) and required official confirmation. Any force majeure clause lasting longer than eight months permitted the cancellation of the agreement.

All three agreements included a provision listing the proper addresses for both Clark and Universal Tractor for all official notices and correspondence. CEAG was responsible for supplying adequate information and training so that UT would be able to produce equivalent products. Should CEAG's technical information be due to failure on the part of CEAG in this regard, CEAG was to correct the defect within 60 days from the official claim. Further, UT could be reimbursed for damages in such an instance through arbitration.

UT was to warranty any product sold back to CIMSA for 1,000 hours of service, or for 6 months from its sale to the ultimate user, or for up to 18 months after shipment by UT, whichever occurred first. CIMSA agreed to provide recommendations as to service parts or repair manuals.

CEAG would not be held at fault for any liability or damages resulting from the designated products (with the exception of faulty technical information furnished by CEAG), but CEAG agreed to assist UT in the defense and settlement of any claims and to help UT in changing the products or method of manufacture to avoid future liability.

Disclaimers were included on both sides wherein neither party assumed responsibility and was not to be held liable for death or any injury or accident that might befall any representative of the other party.

Project Assessment

Factors Expected to Make the Project Easier than Usual

Clark felt that the company's growing familiarity with the Eastern bloc would smooth the way for the project in Romania, at least in part. The fact that Clark did not have to build the new manufacturing facility—but was only to suggest equipment and layout and help acquire items whenever possible—relieved Clark from a heavy load of responsibility while yet maintaining high corporate involvement.

Rapport was expected to be high in that Clark had provided its latest transmission technology, its trademark, and had stated its intentions of purchasing the final product for its own marketing needs.

Factors Expected to Make the Project More Difficult than Usual

Clark did not believe any major difficulties would appear, though it realized that the arrangement with Universal Tractor was the first of this type between Clark and Romania and that each country or new endeavor would include some unexpected problems or occurrences.

Expectations for the Future

Clark's contract with UT in Romania marked another step in the company's penetration of the Soviet bloc market. Clark hoped to increase the trade of its products between Eastern European nations as well as create various sources of its products throughout those countries. It also hoped that each agreement with an individual country would be followed by additional contacts and extended involvements.

HUNGARY

International Harvester Company

Description

The success and growing involvement of International Harvester (IH) in Poland paved the way for a similar nonequity agreement with the Hungarian Railway Carriage and Machine Works (RABA) in Hungary. The initial agreement was signed in 1976 for tandem axles, but this was subsequently expanded to include additional product lines that were added in 1977 and 1978. Unlike the Bumar arrangements concerning construction equipment, the product lines in Hungary involved IH axles and agricultural implements.

Background

Domestic Company. The IH philosophy regarding nonequity transfers apparently changed since its initial agreement with Poland, as the company's desire to enter into such a nonequity arrangement with Hungary was far more positive than in its initial Polish venture. This appeared to be due to two factors. First, IH was now somewhat familiar with the nonequity format, realizing the advantages of such an arrangement in providing a potential manufacturing source without an equity investment. Second, IH found that an agreement with an Eastern European country was not only feasible but also desirable in terms of product and component sales.

IH's corporate background in the agricultural equipment area has been fairly extensive. Its agricultural equipment product line constitutes the second largest of its four general lines. The corporation had a record \$2.33 billion in sales in 1977—compared with \$2.70

billion from the truck division, $731 million from construction equipment, and $208 million from solar turbo machinery. Furthermore, this amount from agricultural equipment sales represented a 3.2 percent increase over the previous year, according to the IH annual report for 1977—at a time when conditions were adversely affecting farming in most areas of the world.

While IH did not have a monopoly in the agricultural equipment area, it did carry a line of farm tractors that were not available everywhere. For example, IH had a line of large tractors having 150 horsepower or more that were used in only three places in the world: the state farms of Eastern Europe, North America (including Canada), and, to a limited extent, Australia. Farms in other areas, such as Western Europe, were too small to justify the use of such large vehicles.

Hungary expressed an interest in obtaining agricultural products from IH as a result of the company's initial overtures. IH, however, could not set up a competing enterprise for construction machinery in Eastern Europe since this would violate its arrangements with Bumar.

Transfer of Technology

Each agreement between IH and RABA involved a separate item within the truck or agricultural equipment line. Several of these agreements contained the manufacturing technology for a component, such as an axle, but not for the entire vehicle.

Turn-key. None of the agreements between IH and RABA included a turn-key approach, as all construction and equipping was done by the Hungarians. IH did not even become involved with the equipment selection, though it did provide suggestions on machine retooling when needed to adjust to the IH technology. Technologies relating to products and processes were transferred.

Know-how. The agreement signed in 1976 between IH and RABA dealt with the IH tandem axle; it was scheduled to conclude in 1986. The 1977 agreement between IH and RABA involved plows and disc harrows. It was also a ten-year agreement and would end in 1987. The 1978 agreement dealt with the IH cycloplanter and was scheduled to run until 1988. After that time, RABA agreed not to disclose the transferred know-how for 20 years. Should RABA desire updates of this know-how after 1988, the agreement would have to be extended.

Each agreement between IH and Hungary was comprised of several contracts and included an open counterpurchase agreement wherein IH agreed to make purchases of a particular dollar amount—

but with no specifications apart from the stipulation that they be something other than farm implements. The agricultural implements agreement, signed on March 9, 1977, included three types of contracts: a license contract, a purchase contract, and contract for the use of IH trademarks on the agricultural equipment. The agreement will run for 10 years, or until January 31, 1987, unless extensions are granted. Four items of agricultural equipment were involved: two models of plows (720 and 800 moldboard plow series) and two of harrows (490 tandem disc harrow series and the 770 offset disc harrow series).

The trademark agreement, effective as of March 11, 1977, included the following categories covering the plows and disc harrows:

1. Official permission
2. Proper use
3. Quality control measures
4. Warranty limitations
5. Ownership
6. Term length
7. Effectiveness
8. Notice
9. Relation to entire agreement
10. Arbitration and governing laws and language

Appropriate exhibits accompanied the body of the contracts, including such aspects as detailed descriptions of the products involved, a statement of the licensed trademarks with appropriate IH logo designs, countries excluded from sales of axles, countries excluded from sales of agricultural equipment, a list of standard parts, and a statement of in-plant inspection and audit procedures.

The sale of axle technology to Hungary included the usual IH promise of updated and improved technology. However, it is relevant to point out that this update technically refers to refinements on the basic product, such as the axle. Should IH come out with a drastically new form of axle, this technology would not be passed along without additional remuneration. It is highly possible that, in such an event, new negotiations might commence regarding the improved axle, particularly if the product comprised an integral part of the IH equipment.

General Management. No management was included in the transfers, but IH did anticipate bringing over its technical people to show how to manufacture the items and to explain why things were done in particular ways.

Patents. The transfer included all necessary patents.

Trademarks. The use of IH trademarks was included for certain products or components meeting quality control specifications.

Buy-back. The buy-back stipulation was essential, although it took the form of a counterpurchase arrangement. A price was specified through a particular date, after which it was subject to increases or decreases depending on the current wholesale price of motor trucks.

International Harvester viewed a buy-back as a situation in which a product directly related to the cooperative agreement is repurchased; on the other hand, a counterpurchase exists when a product unrelated to the cooperative agreement is purchased with the hope that the firm will be able to use the product. For example, IH considered purchasing axles from RABA, since Hungary manufactured 30-ton axles with a reputation for high quality. These axles would not contain IH technology but would be able to be used by IH. If these axles were bought, IH would purchase them in the same way that axles would be purchased from any unrelated company or source.

A barter type of exchange is another option. This occurs when a company does not intend to use the product it purchased but will resell or trade it for an item that the company will be able to use. So far, IH has not used the barter arrangement in Hungary.

Should Hungarian manufacturing of IH farm implements fall behind schedule, such implements could be purchased from IH to satisfy the immediate demand, but IH noted that it did not have a supply of extra axles on hand and would be unable to supplement production in this area.

Other Aspects. The agreements identified marketing territories and listed areas where IH alone would be able to market some of the products. These designated areas varied with the different products in the agreement. Such a designation provided RABA with the opportunity to market the axles to all other areas of the world without infringement concerns.

Responsibilities

Domestic Company. Briefly, the agreements categorized the responsibilities of IH as providing its Hungarian partner with the appropriate technology and accompanying training or instruction and, when authorized, with the IH trademark after all qualifications have been met.

Host. RABA, on the other hand, was responsible for all construction, equipping (when necessary), raw materials, component parts, and for providing employees capable of mastering the skills and techniques required to produce the products involved in the transfer. RABA would also reimburse IH at stated times for the technology. This was to be done through the payment of an initial disclosure fee and subsequent payments, as necessary, for any royalties or research, engineering, and manufacturing (REM) fees when incurred.

Corporate Perception of Benefits to Be Gained from Agreement

Overall, International Harvester felt that its arrangement in Hungary, which was similar to its nonequity agreement in Poland, would complement its activities there as well as increasing IH's reputation throughout Eastern Europe. Furthermore, since Hungary was interested in axles and farm implements, IH's product base throughout the Soviet bloc would be substantially broadened. There was even the potential for an exchange arrangement between the construction equipment being produced in Poland and the axles and farm implements being assembled in Hungary.

In a more immediate sense, IH desired these agreements with Hungary because the corporation would be receiving payments for some of its technology, as well as cultivating another source of axles —either of its own or of current Hungarian design—to supplement its other supplies. However, the additional source of complete products that IH gained through its Polish agreement would not be duplicated in the Hungarian axle project, because IH was not in the business of selling axles, only of using this product as a component in its vehicles.

In Poland, IH felt that some overseas production facilities had to be established to retain its market share in Europe, due to impending competition from five or six other firms. The Hungarian project, however, did not include such a justification for the technology sale, since no competition was perceived. As with Poland, however, IH foresaw an increased market potential in an area where it previously had no market share.

Negotiations

Parties Involved

The agreements were all signed by International Harvester Company (the domestic company) and the Hungarian Railway Carriage and Machine Works (RABA).

Initiating Contact

Unlike the arrangement between IH and Bumar, in which Poland first sought out the U.S. company, IH was the instigator in the Hungarian projects.

Length of Negotiations

The first contact between IH and Hungary regarding these projects occurred in 1970, but serious discussions did not take place until the fall of 1975; the first agreement was signed in 1976.

Issues and Obstacles

The most intriguing aspect of the Hungarian arrangement was the counterpurchase issue in which IH agreed to a certain monetary amount of purchases. This remuneration then provided Hungary with the funds necessary to buy essential components from IH to make its manufacturing and assembly ventures possible in the farm implements line.

Time Frame

Overview

The 1976 tandem axle agreement will run until 1986; the 1977 plow and harrow agreement will conclude in 1987; and the cycloplanter agreement of 1978 concludes ten years later, in 1988.

Financial Flows

Payments

In this reciprocal type of arrangement, IH's payments were tied indirectly to RABA's production of Hungarian-designed axles—not to production of the IH products. As in the Polish project, two types of payments were involved.

Purchases from the Domestic Company

IH agreed to a counterpurchase of a certain amount of axles, assuming those axles were available. Prices for those axles were firm through a certain date but were subject to a percentage change based on the wholesale price of truck parts after that date through the remainder of the agreement.

Purchases by the Domestic Company

No purchases of any consequential size were anticipated on the part of IH.

Personnel

Training Tasks

As in its other technology sales, IH agreed to provide training both in the United States and in Hungary for RABA employees involved with the transfer. No problems were anticipated, particularly because IH then had prior experience in such nonequity technology transfers and because, in the case of Hungary, RABA employees were already manufacturing axles (although not of the IH design).

Other Personnel Involvement

No other involvement was anticipated.

Protection

Leverage

Domestic Company. Throughout the negotiations with Hungary, IH maintained a relatively strong position. Hungarian dependency upon IH purchases provided both partners with a strong incentive to work together.

To ensure that proper manufacturing methods were used, and also that the final product was according to specifications, IH agreed to test the completed components. If the product did not meet IH's standards, it was to be sent back. This arrangement provided IH with additional leverage, since the Hungarian operation had supplied the funds for this transfer and would not be reimbursed until IH repurchased the axles. Thus, RABA could be out $20 million to $30 million at some point, which would delay other programs that depended on the availability of funds.

Host. Hungary maintained the usual leverage in that the country offered an extended market to IH as well as immediate payment for aspects of IH technology. In this case, the contracts reflected all guarantees in the form of responsibilities on the part of each partner, supported by financial payments by RABA or product purchases by IH.

Inspection Privileges

Inspection provisions included IH's right to quality test all products according to predetermined specifications. IH was given the right of inspection throughout the duration of the contract.

Patent/Trademark Defense

If a patent acquired from IH should be challenged, IH agreed to defend RABA's position at IH's cost. Under U.S. law, IH has the right to grant patents to others. Permission to use IH trademarks was included in the agreement, providing all tests and standards proved acceptable.

Legal Provisions

All IH contracts with Hungary carried the same arbitration clauses and referred to the Swiss arbitration court in Zurich should difficulties or disputes arise. A force majeure clause is also a permanent clause in IH contracts of this sort.

Project Assessment

Factors Expected to Make the Project Easier than Usual

International Harvester expected the arrangements with Hungary to be easier than with Poland for three general reasons. First, the magnitude of the project was much smaller than the manufacturing venture with Bumar for construction equipment; hence, it should be less complicated. Second, IH felt Hungarian workers with RABA would require less training because the technology to be transferred was not as extensive and because their prior experience was more appropriate to the project. Third, as a result of its experience in Poland, IH had a body of experience with nonequity technological transfers upon which to draw in effecting subsequent arrangements.

Factors Expected to Make the Project More Difficult than Usual

The interdependence between different aspects of the IH-RABA agreements was a recognized concern, as the payments for farm implement components were tied to IH purchases of RABA products, which in turn made RABA dependent on its own production schedule. This placed IH in the middle. Apart from this, there were the usual concerns accompanying most technological transfers on an international scale, but no overt problems were anticipated.

Expectations for the Future

International Harvester hoped to continue to expand its production sources as well as product marketing throughout Eastern Europe, and the Hungarian ventures were steps in these directions. In its immediate plan, IH hoped to expand the types of operations and technology transfers within Hungary while keeping negotiating channels open with the rest of Eastern Europe.

YUGOSLAVIA

The Bendix Corporation

Description

At the time of this writing, The Bendix Corporation, headquartered in Southfield, Michigan, was involved with several nonequity agreements with countries in the Soviet bloc.

The agreement examined here is a License and Technical Assistance Agreement between The Bendix Corporation and Varazeinski Tekstini Komdinat (Vartek) in Varazdin, Yugoslavia, for the manufacture of organic friction materials for disc brake pads for a specific model of automobile produced in Yugoslavia. The agreement was validated by the Yugoslav government on January 24, 1975, and extends through January 24, 1985.

Background

Domestic Company. The Bendix Corporation's involvement with countries in the socialist sphere, such as in Eastern Europe and the USSR, demonstrates the corporation's lack of problems with past and present agreements. It also validates the corporation's assessment that an ongoing market exists in those countries for the types of products that Bendix can supply.

The agreement between The Bendix Corporation and Vartek in Yugoslavia came about through the involvement of the corporation's French subsidiary, Ducellier Bendix Airequipment, Societe Anonyme (DBA, S.A.)—owned 92 percent by Bendix, U.S.A. This is not the only contact between Bendix (Southfield) and Eastern Europe or the USSR, however. A License and Technical Data Agreement between The Bendix Corporation and Licensingtorg of the USSR was signed in March 1977. The licensed product covered under the terms of that agreement was a power booster for the vacuum power brakes for a certain model of automobile. An earlier agreement, signed in 1968, involved another model with another variation of the brakes. The 1968 agreement provided the technical assistance required to produce the brakes; therefore, such assistance was not a requirement of the 1977 agreement.

The Soviet agreements also included licenses for the applicable patents for the products. Both the 1968 and the 1977 agreements provided for the payment of a one-time fee for product technology, with no training or process information provided and with no subsequent revenues from product sales required to be paid back to Bendix.

Various subsidiaries of The Bendix Corporation, especially those headquartered in Western Europe, also have contracts of a nonequity nature with either Eastern Europe or the USSR, but the above agreements are the only ones held by the U.S.-based corporation as of this writing.

Perhaps the most advertised arrangement between Bendix and the USSR was the tentative plan for the construction and operation of a spark plug plant within the USSR. Despite premature descriptions of the agreement, released in January 1977, the contract talks are now, as described by one corporate official, "placed on ice indefinitely." However, there are several other agreements in the final stages of approval, including one for industrial gauges between Bendix and Licensingtorg of the USSR, which includes both a license and a technical assistance agreement. Another venture, in the planning stage, is between Bendix and Hungary. Negotiations for both are in a nonequity format and involve the transfer of technical data.

Host. The signed agreement between Bendix and the Vartek company involved the manufacture of friction material for disc brakes. The factory is located in the town of Varazdin, population 40,000; over 10,000 of its citizens are directly employed by the Vartek company. The agreement was concluded with Bendix, although only one of the divisions, the friction material division, was to be directly involved with the supply of technology. Yugoslavia manufactures its own cars through a license agreement with Fiat.

Transfer of Technology

The agreement examined here will concentrate on the technology transferred from the U.S.-based corporation to Yugoslavia. Bendix was to furnish technical assistance and initial consultation for the layout of the plant and the production equipment, production equipment specifications (including all raw materials specifications), and assistance in implementing and maintaining production of the licensed products.

Turn-key. The extent of Bendix's responsibility did not extend to the construction or equipping of the factory, but it did include a list of suggested equipment to be purchased, including alternative purchase items when possible.

Information and Know-how. All technical information related to the preparation of the friction material was included in the transfer of the product's technology, while the transfer of the actual manufacturing process would be accomplished through training sessions held both on site in Yugoslavia and in the United States at the Bendix plant in Troy, New York. The information was transferred in written English as well as through oral instructions. Information as to improvements will be updated until January 1983. The technology transferred applied strictly on one vehicle model, such as a Fiat 125 or 126. Should Yugoslavia desire to use this product on a different car model, a formal letter of request must be made and approved.

Management. No specific managerial techniques were included apart from production know-how.

Patents. A patent agreement was included in the contract, although its importance was apparently in its application to third-party infringements; Bendix did not carry a patent on the friction formula in Yugoslavia. When the patent expired it would also expire with regard to the agreement. At that point, the patent would fall into the public domain.

Patents related to improvements or updating of the friction material must be submitted to Yugoslavia within eight months after the filing date. If Yugoslavia should elect to keep the patent, it must pay the submission fees to file it in that country. If Yugoslavia does not accept the patent within 60 days, Bendix is free to offer the patent to other potential licensees. This freedom actually exists whether or not the country accepts the patent, since normally all licenses are granted on a nonexclusive basis.

According to the terms of the agreement, the purchaser—in this case, Vartek—is responsible for paying the filing and prosecution expenses for application of relevant patents in its specific territory. The purchaser must also reimburse the seller for any expenses incurred in such patent applications. Should Vartek fail to maintain the necessary patents or not accept the initial patent application within 60 days of its filing, it would renounce all rights to those patents. Bendix also disclaimed all responsibility for the product manufactured in Yugoslavia regarding its sale or utilization.

Trademarks. No trademark rights as such were included in the transfer agreement, although Vartek is permitted to state, "manufactured under license by Bendix." It has been general corporate policy not to permit unaffiliated companies to use the Bendix trademark except for exceptional circumstances; this agreement was not exceptional.

After an agreement expires, all rights and privileges bestowed by that agreement normally dissolve—including the use of all trademarks or corporate associations. Such is the case in this instance, unless the agreement is extended. It is highly probable that the agreement may be extended beyond 1985, including all technological updates relating to the friction materials. In such a case, the extended use of the Bendix manufacturing license logo would be most likely.

Buy-back. Automobiles manufactured in the United States do not use the friction material specified in this agreement; hence, no buy-back clause was included, even though such a clause is often part of non-equity agreements with countries in Eastern Europe. All purchase fees, with the exception of the initial disclosure fee, were based on a percentage of sales, with a minimum royalty fee applied annually. Yugoslavia was not to export the friction material alone, although the product certainly could be exported as original equipment on a complete automobile.

Other Aspects. Bendix did have the option of transforming this non-equity agreement into that of a joint venture. This would have been feasible because Yugoslavia was one of the few countries in Eastern Europe that permitted such an alliance with a capitalist country. However this option was only viable for the first five years that the agreement was in effect, or until January 24, 1980, and was not part of the official agreement.

Responsibilities

Domestic Company. Bendix agreed to provide technical documentation relating to the plant, the production, and the formulation and processes necessary to produce the organic friction materials. The corporation furnished a suggested listing of necessary equipment required for such manufacture and offered requisite training toward the production of the friction compound. The list of equipment suggestions included raw materials sources in addition to the equipment itself. In some cases, materials would initially be imported from Bendix or alternative sources, with a gradual switch to Yugoslavian materials if available. Personnel training was a two-part endeavor: skilled technicians from Bendix would travel to Yugoslavia on a visitation basis, and technicians from Vartek would attend training sessions in the United States. Since no construction or long-term training was required of Bendix, there were no provisions for lengthy housing or transportation of families to Yugoslavia from the Bendix offices. Bendix was expected to perform quality control checks on the product, even though the item would not carry the Bendix trademark, because certain specifications were required to meet the standards of the original equipment manufacturers (OEMs). The OEM in this instance was Zestava, a Yugoslav company marketing automobiles under license from Fiat. The OEM would also perform its own tests on the product, since some of its standards might vary from those of Bendix. Thus, while Bendix approval was not the only control, it was an important one. Producing quality materials within Yugoslavia was also important in terms of the overall product, since it was hoped that approximately 95 percent of the parts for each car manufactured there would be locally produced—including the brake mechanisms in the near future.

Host. Vartek's responsibilities in this agreement consisted of paying the assigned royalties. The quarterly royalty payments were to be accompanied by royalty reports, which listed the products sold, the turnover obtained by Vartek through such sales, and the amount to be paid to Bendix according to the agreement. Payment of at least a minimum fee was mandatory, regardless of manufacturing delays; therefore, it was to Vartek's benefit that the factory be equipped, staffed, and functioning in the shortest possible time so that sales of the product could be used to provide funds for the royalty fees. Vartek was also responsible for all travel expenses involved in the operation of the agreement.

Corporate Perception of Benefits to Be Gained from Agreement

There were two major reasons underlying The Bendix Corporation's desire to undertake the project in Yugoslavia. The foremost reason was to take advantage of available technology that was not being utilized at that time in Eastern Europe. The second reason was to gain entry to a market area—specifically that of Yugoslavia, but generally that of Eastern Europe.

Negotiations

Parties Involved

Negotiations were handled by the International Division of Bendix through the vice-president of licensing, whose office handles licensing around the world for Bendix in all four of its manufacturing areas: aerospace/electronics, automotive, industrial, and forest products. Negotiations with company officials of the Yugoslav factory were carried out directly, with no interference from governmental officials.

Initiating Contact

Braking components were initially supplied through a Fiat-operated company, but Yugoslavia desired local production of the braking components and approached the Bendix subsidiary, DBA, in France for the technology to produce the disc brake. Vartek's request for the formula for the friction compound was combined with a request for the disc brakes, but DBA did not possess the friction material information. However, DBA's parent company, The Bendix Corporation, in Southfield, Michigan, did have this technology. It is interesting to note that, while The Bendix Corporation in Southfield had this information, the corporation was not then using it, because the braking components of automobiles in the United States required the use of a different process. Thus, this sale provided a use for the technology without creating a competitor for Bendix, at least in the domestic sense.

Separate agreements were drawn up between Yugoslavia and DBA in France for the manufacturing technology for the disc brakes and between Yugoslavia and Bendix in the United States for the friction materials.

Length of Negotiations

The first contract was drawn up by December 1973, but over a year passed before the agreement was officially ratified in Yugoslavia on January 24, 1975.

Issues and Obstacles

No particular obstacles emerged during the negotiations. Yugoslavia was desirous of producing the product within specific quality guidelines. Bendix was concerned that the product be produced in good quality. When assured that Yugoslavia would honor this desire, no stumbling blocks remained.

Yugoslavia was one of the few Eastern European countries that permitted equity ventures by foreign corporations with up to 40 percent ownership; but, in this case, Bendix did not want such an arrangement. However, as mentioned earlier, the possibility of expanding the agreement to that of a joint venture at a later time remains open if such an arrangement is desired.

The option to transform this nonequity arrangement into a joint venture appeared suitable to both parties within the five-year time frame, although the desire to invest in the project appeared to have been originally suggested by Bendix.

Time Frame

Overview

No time frame detailing activity between the opening and closing dates was included in the agreement, although certain time limits were established, for example, for the patent acceptance period. The joint venture option might also be viewed as a time constraint, although this was not part of the license agreement. Following the initial ten-year period, the License and Technical Assistance Agreement would continue on from year to year unless a written notice of intention to cancel was provided by one of the parties.

Special Considerations

An opening issue in the agreement notes that should Yugoslavia not ratify the agreement within three months after the parties have indicated their approval, the parties would sit down together to work out areas of dissension or to decide how to proceed. This issue regarding validation of the contract did not arise, however, because Yugoslavia did ratify the agreement.

Production was expected to begin within three years of the effective date of the agreement, or by mid-1978, although a date was not specifically mentioned. Several other time limits were not detailed in the agreement; for example, there was no time limit placed on training periods. Instead, emphasis was put on the mastery of certain techniques, as determined by the trainers.

In the case of default by one party of any of the essential obligations of the agreement, the other party had the right to terminate the agreement 60 days following its notice to this effect. However, should the default be remedied within the specified period, termination would not ensue.

Financial Flows

Payments

Neither party was permitted to give out the financial amounts of the contract in detail. However, a general approximation of the size of the agreement places the initial disclosure fee in the range of $100,000 to $500,000. Assuming minimum royalty payments over the course of the remaining nine years of the life of the current agreement, it is expected that the total amount paid to Bendix by Yugoslavia will be approximately $500,000 to $1 million. Of course, this amount would increase should production rise, and the minimum royalty payment would be proportionately increased.

An initial disclosure fee was paid at the start of the agreement, and subsequent payments would be made on an annual basis thereafter for the duration of the nine remaining years of the contract. Until production was in force, these payments would be at a minimum level. Following start-up, royalty payments would increase, if necessary, because they were based on a percentage of gross sales. All payments would be in U.S. dollars guaranteed by an acceptable bank, with no inflation factor but also with no discounts included. The use of the gross sale price, however, would take price fluctuations at least partially into account. The payments would be guaranteed by a Yugoslavian bank and placed in a locked box bearing the address of Bendix.

Because royalty payments were based on a percentage of gross sales, Bendix maintained the right to obtain all information on Vartek's sales of brake compound friction materials that would lead to a correct audit. Because the material would be used in the construction of cars, the number of automobiles produced in Yugoslavia also provided a backup check for approximate sales of the item.

All royalties not paid within 30 days after each quarter of the calendar year would be assessed a 10 percent interest charge. Bendix was permitted to examine all records related to production and sale of the product at reasonable intervals for up to three years following the conclusion of the agreement.

Interestingly, Vartek was obligated to pay any income taxes assessed against Bendix by Yugoslavia (up to but no greater than the amount of tax assessed against Bendix relating to this product in the United States). These payments would then be deducted from the amount owed to Bendix under the determined disclosure fee and subsequent royalties stated in the agreement.

Purchases from the Domestic Company

To aid in its personnel training, Vartek purchased some completely premixed compounds of the friction materials from Bendix at a price less than the manufactured items. These were purchased several months prior to the anticipated start-up date of January 1979 and were intended to provide functional training more than to aid the factory's inventory level. Bendix did not expect Vartek to purchase finished items of the friction materials as a carry-over until their own production was under way, although Vartek did have a contract to provide the product to the Yugoslav car manufacturers.

Vartek was responsible for all travel expenses incurred on the project, either by Bendix or Vartek employees. Furthermore, Vartek was to reimburse Bendix for any and all expenses incurred in the transfer of technology to Yugoslavia over and above all travel expenses. This reimbursement was to be made within 30 days following receipt of the invoice for such expenses.

Personnel

Approximately 200 Yugoslavians were expected to be involved in production of the brake friction materials, but this estimate included both the French subsidiary's project as well as the U.S.-based one. As stated before, Vartek was responsible for all travel expenses.

Low employee morale on the part of workers involved with the Vartek project was not viewed as a potential problem. Some of this optimism stemmed from Vartek's very prestigious reputation, at least in the area of textiles, and some from the company's active role in community affairs and in the town itself. Production problems were expected to be slight and would probably be due to inexperience, since the production of the friction materials varies greatly from the production of textiles there.

The agreement listed no skill areas required of selected Vartek workers. Furthermore, no training schedule was included, and the number of workers instructed in the United States or the quantity of man-hours or man-days was not documented. All training provided by Bendix in the United States would be conducted in the Bendix facility located in Troy, New York. However, some additional training was also to be given at the Bendix subsidiary located in Spain, an affiliate skilled in this form of friction materials manufacture.

Protection

Leverage

Domestic Company. The leverage possessed by Bendix throughout the agreement depended greatly on the value placed by Vartek on the friction material, and on the personal relationship that developed between Bendix and Vartek. Any Bendix improvements in the material after 1983 (when the technology was no longer being transmitted to Vartek) should only increase the desirability of the project's extension.

Host. The same leverages were also held by Vartek, since Bendix had indicated its desire for future transactions. Thus, both partners were in good positions to mutually satisfy each other's wishes for additional involvement.

Inspection Privileges

Inspections or quality control checks were permitted by both sides. Bendix would use the inspections to track the quality of the product as a method of ensuring correct transfer of the technology. The host and the OEM would use their own quality control checks to approve the product before its installation on the automobile.

Patent/Trademark Defense

In the event of a threatened infringement suit against Vartek or Vartek's customers, Bendix agreed to freely provide Vartek with any information in its possession that would assist Vartek in dealing with the suit.

A reciprocal patent opportunity was included in the event that Vartek improved upon the original technology provided by Bendix. In such an event, Bendix would then be granted a nonexclusive and nontransferrable license to manufacture and sell the improved products throughout the entire world.

Legal Provisions

Bendix approved exports of the brake friction material to other countries only within the Eastern European bloc. Vartek was allowed, however, to export licensed products as part of the original equipment on a complete vehicle—in this case, the automobile. Vartek was also permitted to provide replacement parts for all vehicles carrying this particular friction material as part of the vehicles original equipment. This permission covered not only those vehicles sold in Yugoslavia but also those exported to any other country.

The technology transmitted to Vartek was the state of the formula as of 1975, although all updated improvements would be provided to Vartek until two years before the completion of the agreement, or up to January 1983.

Bendix described the relationship as one of assistance and helpfulness; if Bendix was able to help its Yugoslav partner, it would do so. Bendix viewed the project more as a long-term relationship—that might go beyond the ten-year contract date in either an extended or expanded measure—than as a terminal project for only immediate gain.

Protection aspects written into the contract included an arbitration clause referring settlement to an outside court located in France, subject to the Rules of the International Chamber of Commerce. Interestingly, there apparently was no force majeure clause. However, Bendix was permitted to terminate the agreement in the event of certain circumstances or actions on the part of Vartek—such as an unjustified delay of payment; unauthorized use of the Bendix trademark; failure to respect specified product standards relating to quality, insolvency, or bankruptcy; or a total cessation of activity.

Project Assessment

Factors Expected to Make the Project Easier than Usual

Both Bendix's and Vartek's attitudes toward this project comprised the starting point for a much longer and more fruitful relationship; they were considered a positive factor in smoothing out any problems that might arise during this particular agreement.

Factors Expected to Make the Project More Difficult than Usual

Nothing was anticipated that might make the agreement more difficult than usual.

Expectations for the Future

The future was Bendix's major consideration in this contract, because Bendix desired to continue its association with Vartek, in particular, and with Yugoslavia, in general—even expanding its contacts throughout Eastern Europe. The immediate option of establishing a joint venture with Vartek aided this desire, as did the possibility for extension of the agreement beyond 1985.

NOTES

1. Eric W. Hayden, Technology Transfer to East Europe: U.S. Corporate Experience (New York: Praeger, 1976), p. 45.
2. Ibid., pp. 49-50.
3. Ibid., p. 46.
4. Ibid.
5. Ibid.
6. Ibid., p. 48.
7. Ibid., p. 50.
8. Ibid., pp. 46-47.
9. Ibid., pp. 48-49.
10. Ibid., pp. 49-50.
11. Ibid., p. 54.
12. International Harvester Company, Annual Report (Chicago, 1977), p. 2.
13. Jerry Szumski and Janusz Piotrowski, "Bumar (Poland)—International Harvester (U.S.A.) Cooperation Is Now Over Five Years Old and Keeps Going," Office of Polish Commercial Consul (Chicago, May 1978).
14. Polish Machine Industry Offers 5 (73) (1979): 4.
15. Szumski and Piotrowski, "Bumar," p. 7.
16. Polish, p. 5.
17. Szumski and Piotrowski, "Bumar," p. 8.
18. Ibid.
19. Polish, p. 8.
20. Szumski and Piotrowski, "Bumar," p. 9.
21. Polish, p. 4.
22. Szumski and Piotrowski, "Bumar," p. 10.
23. Ibid.
24. Ibid.
25. Hayden, Technology Transfer, p. 54.
26. Szumski and Piotrowski, "Bumar," p. 3.
27. Ibid.
28. Ibid., p. 10.
29. Ibid., p. 4.
30. Hayden, Technology Transfer, p. 95.
31. Ibid., p. 25.
32. Ibid., p. 59.
33. Ibid., p. 62.
34. Ibid., p. 66.
35. Howard Wolff, "Polish Plant Ready to Roll," Electronics, October 14, 1976, p. 75.
36. Hayden, Technology Transfer, p. 65.
37. Ibid.

38. Ibid.
39. Ibid., p. 68.
40. Ibid., p. 57.
41. Ibid., p. 66.
42. Wolff, "Polish Plant," p. 74.
43. "Westinghouse Gets Pact to Build Plant in Poland," The Latrobe (Penn.) Bulletin 11 (December 1973): 1.
44. "Westinghouse Turnkey Contract Typifies Poland's Need for Top Flight Technology," Business International 20 (December 14, 1973): 393-94.
45. Hayden, Technology Transfer, p. 68.
46. The main contract for equipment purchases was for $2.5 million, but add-ons brought this amount up to a total of $3.7 million.
47. "Westinghouse Turnkey Contract," p. 394.
48. Wolff, "Polish Plant," p. 74.
49. Gus Prekopa, "Negotiating Technology in Romania," Les Nouvelles, Journal of the Licensing Executives Society 10 (June 1975): 86.
50. Hayden, Technology Transfer, p. 92.
51. Ibid.
52. Prekopa, "Negotiating Technology," p. 86.
53. Hayden, Technology Transfer, p. 93.
54. Prekopa, "Negotiating Technology," p. 86.
55. Prekopa, panel presentation at the 75th Annual Meeting of the National Machine Tool Builder's Association, October 29, 1976, p. 4.
56. Prekopa, "Negotiating Technology," p. 86.
57. Ibid., pp. 86-87.
58. Hayden, Technology Transfer, pp. 93-94.
59. Ibid., p. 94.
60. Ibid., p. 92.
61. Ibid.

4

DESCRIPTION AND ANALYSIS OF CORPORATE OUTCOMES

INTRODUCTION

This chapter follows the same general format as the preceding one but concentrates on the outcomes of those nonequity arrangements between selected corporations and countries within Eastern Europe. As before, each case is handled separately, but the same outline has been used for all eight projects.

Outline

The outline used for detailing the outcomes of these nonequity arrangements follows the same structure as the one used for describing expectations.

COUNTRY

Company

Description variations

Current corporate perception of benefits from agreement

Negotiations
- Contract format revisions
- Pending negotiations

Time frame
- On schedule
- Behind schedule
- Other aspects

Financial flows
- Payments
- Purchases from the domestic company
- Purchases by the domestic company

Personnel
- Training assessment
- Other personnel involvement

Protection
- Leverage assessment
- Utilized inspections
- Patent/trademark defense
- Legal provisions invoked

Hindsight
- Factors that made the project easier than anticipated
- Factors that made the project more difficult than anticipated

Progress of future expectations

Time Qualifications

It should be pointed out that most of the agreements discussed here were still in progress at the time information was obtained. (In fact, the only agreement that had officially concluded was between Westinghouse and Poland. Honeywell's arrangement with Poland was due to finish by 1979 but an extension was known to be in the offing.) Therefore, this summary of corporation outcomes is based upon perceptions up to a certain point in time. In some instances, it is highly possible that these outcomes will vary or even change drastically before the agreements are actually completed.

POLAND

Clark Equipment Company

Description Variations

The 1972 agreement between Clark Equipment Company (with its foreign subsidiaries) and Bumar Foreign Trade Enterprise involved the manufacture of Clark-designed axles in Bumar plants located in Stalowa Wola and Radomsko. The output was intended to supply the needs of both Clark's and Bumar's customers. Five years after the instigation of this arrangement, one of Clark's European subsidiaries reported using the Polish-built axles "with excellent results,"[1] a tribute to the functioning of the project. However, axles from the Radomsko and Stalowa Wola plants were manufactured at a low production level. No new Clark axle models had been introduced in 1977. This was attributed to a relatively flat market and, to a lesser extent, a lack of hard currency. However, one additional Clark model was slated for release by Bumar in 1978.

Clark officials reported that the Bumar factories were being used as axle sources, with 257 axles purchased from Bumar through Clark International Marketing S.A. (CIMSA) to meet a Clark demand in France in 1977. Bumar had been reliable regarding delivery time, product reliability, and axle quality, Clark reported.

Clark officials noted that, at times, dealing with the Eastern Europeans was a far more satisfactory experience than similar contracts or arrangements with other groups or regions. Their experience has been that the Eastern European partners hold to their end of the bargain, that they live up to the letter of the contract. Clark officials observed that their Eastern European partners had liked their U.S. partners, trusted them, and heeded their instructions. With this chemistry at work, Clark felt that similar arrangements would have a better-than-average chance of working. Clark also felt that such contractual arrangements should be relatively ongoing, with updates and the need for improvements worked out during the initial time period. Clark felt that a technology transfer over a stated time period provided a high level of support that could transform the transfer into more of an extended operation.

Joint Project

Bumar possessed a license from Clark Equipment for the manufacture of axles. However, what might have been described as a "rather classical license agreement" was extended to cover several new variations of axles designed entirely by Polish specialists. Technical solutions were fully approved by Clark, and the new axles are now being produced and sold as jointly developed, Clark-Bumar products.[2]

Clark was very pleased with the metric axles; their production should benefit both Clark and Bumar. The dead-steering axle amendment was signed on October 25, 1975.

Perhaps one of the most unexpected occurrences during the agreement between Clark and Bumar took place when the partners joined talents to produce an excavator axle that Bumar had previously been purchasing from the USSR. Clark felt that this joint venture promised a "new way for us to go" in its international dealings. This venture, at least, should be successful, since the demand already exists for the product.

The amendment to design an axle with metric dimensions that would satisfy Soviet demand was signed in April 1975. It called for the joint development of a hydraulic excavator model to replace the present Bumar axle used in the USSR. Bumar had been exporting axles to the USSR for the past six or seven years, but the design was obsolete. Bumar embarked on the joint development project with Clark to create an updated axle of metric dimensions to improve this design.

Another benefit resulting from the know-how exchange between Bumar and Clark was the extraneous information gained by Clark concerning the approach taken by the USSR in many of its construction projects. For example, during the joint axle development project, Clark found a distinct difference between the method of operation of the USSR and the United States in pipeline construction. The United States will build a road alongside the pipeline being constructed and bring equipment down as the work progresses. The USSR will fly in an excavator in parts, assemble it for a specific area, but then will just leave it and fly in another vehicle in parts further down the way.

Transfer of Technology

Clark felt that it never had a serious problem with Bumar regarding quality control. Clark maintained the right to inspect all products carrying its trademark, and it held to the standard that no product would be approved that was not up to specifications. Part of the incentive behind the high quality of Bumar's workmanship was explained by one Clark official who stated that Bumar simply wanted to work with Clark. Because of this desire Bumar followed Clark's instructions with great care.

Other Skills Required

Besides the actual product know-how, other skills were transferred as well—cost control, for example. Clark officials remarked that Bumar had become increasingly cost conscious. "Seven years ago," one Clark official noted, "Bumar would not have counted the number of nuts being used—now they are counted carefully."

Inventory Control

Since Clark was involved in a form of buy-back arrangement with Bumar, the corporation wanted certain models manufactured that could then be used to meet its needs elsewhere. However, Clark found it was more difficult than anticipated to arrange to have Bumar manufacture those particular models. Furthermore, the whole area of inventory control was more difficult than expected.

Current Corporate Perception of Benefits from Agreement

The three reasons given for Clark's desire for the agreement with Bumar included a manufacturing source for Clark's international market, especially in Europe; an entry vehicle into Eastern Europe; and the profit gained from the disclosure fee and component sales. The results from the arrangement varied somewhat from the expectations.

Clark's primary justification for the arrangement with Bumar at first had been to establish a source of supply of heavy-duty axles on the European continent because the cost of importing products from Clark, U.S.A., price such units beyond local market prices. However, during the 1978 interviews with Clark officials, this rationale for effecting the technology transfer was described as "limited" at best. This change may have been influenced by the delays that slowed the production rate and hampered the supply of axles.

Clark found that the business cycles that affect the United States are often reflected in Europe six to nine months later. For instance, the general recession that peaked in the United States in 1975 reached a saturation point in Europe during the latter months of 1975 and 1976. This brought about a slowdown in the buying of replacement parts and units, which has only recently resumed its anticipated volume. Thus, sales were less than had been anticipated.

Clark's enthusiasm for this transaction as the corporation's entry point into Eastern Europe, which had been highly optimistic in late 1974, apparently cooled off—perhaps because many of the possible agreements that were being considered at that time did not materialize. By 1978 Clark had adopted a "wait-and-see" approach. On the other hand, more contacts had been made within Poland and throughout Eastern Europe than apparently had been initially expected in 1972.

Clark gained extra "mileage" from its technology in terms of the disclosure fees paid during the agreement. This payment could be used for additional product development as well as for offsetting past development expenses. Clark received only minimum payments for engineering fees and royalties because of delays in the production schedule and the drop in sales. These lower payments were partly offset by additional component sales (which had exceeded expectations) that occurred when the Polish axles had not been manufactured as rapidly as anticipated. These increased sales referred only to components sold to Bumar before its own facility was on-stream. Sales of components incorporated into the final product were not included in this tabulation. In the latter regard, sales were understandably down because this component category was directly tied to production.

An unexpected sales opportunity for products not directly connected to the axles did transpire—Clark's export sales did increase during the course of the company's involvement with Bumar.

Negotiations

The comprehensiveness of the agreement between Clark and Bumar was apparently adequate; it has covered all contingencies that have arisen thus far.

Contract Format Revisions

A side agreement was drawn up that permitted Bumar to use some of the components being produced for other applications. In addition, four amendments were added to the original agreement. The first granted Bumar permission to produce what was termed a dead-steer axle; the second and third amendments were never signed or put into practice; the fourth amendment was designed to expand the agreement into the joint development project creating a metrically scaled hydraulic excavator axle for export to the USSR.

The license for Bumar's manufacture of Clark axles was expanded since the original agreement to "cover new types of products, designed entirely by Polish specialists, whose technical solutions have been fully approved by the firm of Clark Equipment Company, and are now being sold as a jointly developed Clark-Bumar product."[3]

Pending Negotiations

Clark negotiators expected to conclude arrangements based on the reputation of their company's name—not realizing that the individuals involved in the negotiations must become acquainted with their foreign counterparts on a personal basis for the arrangements to be taken seriously and concluded amicably. Clark found that the speedy time involved in settling the contract was not totally advantageous. The negotiating period can be a time of becoming acquainted and of ironing out contract issues. Clark's agreement with Bumar had been concluded so rapidly that this familiarization time occurred afterward, when the project was under way. It would have been advantageous to have this rapport already established prior to setting a project into motion.

There is a tendency for particular individuals to come into prominence as situations create a need or demand for their particular skills and/or background. Clark found this to be true in its negotiations with Bumar. One of the company's best and chief negotiators was a person familiar with Eastern European culture as well as with the industrialized products being discussed. It must be pointed out that much of the smoothness of the negotiations and of the project itself was due to this individual and the rapport that he was able to establish.

Time Frame

On Schedule

Axle production in Poland appears to be on course now, although the variety of axles produced is not as great as some had hoped. Demand, however, does not seem to be as high as initially projected either. This may be due to a large demand for a variety of axle types rather than a high demand for relatively few models.

Behind Schedule

The projected time frame for the Clark-Bumar project ran into several delays. The first major delay was caused by difficulties in obtaining necessary equipment; then, when production was getting on-stream, the 1975 recession that was affecting Europe reduced demand for the products. The initial delays were frustrating to both Clark and Bumar, and since the axle was a vital component in a number of vehicles, it affected Polish production in other areas, too. Many of the axles were needed for cranes being produced in Poland; thus, the shortage of axles hindered the supply of cranes as well as affected the revenues returned to Clark as royalty payments. During some years only 100 to 200 axles were produced—at other times, only 50 axles.

In retrospect, Clark felt that part of the problem with the production quotas was that the tentative schedule listed in the contract reflected the amibitions of Bumar more than the reality of such a technology transfer. Forecasts that were so optimistic could only be disappointing when the goals were not achieved.

One of the most negative results of the production delay was that the entire Bumar crew was fired. This was upsetting, as one Clark individual noted, because these individuals were fired for causes over which they had no control. Clark's impression of its Polish colleagues was that they were very good engineers.

Financial Flows

Payments

Financial flows from Bumar to Clark have taken place generally as anticipated. The first disclosure fee was a bit late owing to a misunderstanding, and some royalty payments were behind schedule, but this is almost considered "normal" worldwide. The lower production levels and the drop in demand for axles, at least in Europe, resulted in lower payments to Clark because the company was then paid only the minimum royalty level each year. However, this is expected to improve as construction and other areas of business that use vehicles with these axles pick up.

Purchases from the Domestic Company

Direct product purchases from the domestic company were higher than anticipated, since some axle components were purchased from Clark because the need was there prior to Poland's production capacity. Other component sales—those assembled into the axles during production—were down because they reflect a direct relationship to production. Purchases throughout the Eastern bloc of Clark products unrelated to axles were up more than originally anticipated.

Purchases by the Domestic Company

Generally, Clark has not needed to purchase any unexpected items for work on this project. This is partly due to the fact that all visits to Poland by Clark personnel have been made only on a visiting basis, so no long-term accommodations have been necessary.

Personnel

Training Assessment

Domestic Company. In 1976 Hayden noted that Clark encountered personnel problems such as a lack of worker training, low quality control, and worker indifference—all contributing to a longer pre-start-up time before production.[4] However, by 1978 the situation had become more positive. Clark noted that it found the technical capability of the Polish engineers outstanding and the manufacturing efforts in the plants of high quality.[5]

Actually Clark found the capability skills within Bumar to be very good—in fact, the supply of academically trained engineers was even better than expected. Manufacturing capabilities appeared to be approximately what Clark had anticipated.

The company felt that Poland's "abundance of well-trained and capable engineers and manufacturing plant workers," combined with the "American companies manufacturing and plant operation know-how, plus an ability to market products in the Western markets . . . seem to be far more important than the low Polish labor costs and highly sophisticated American products which are commonly considered to be the major incentives for mutual action."[6]

After the equipment was installed, Clark sent technicians to Poland at Bumar's request to teach the Poles the operations on their own equipment. No Clark representatives were in Poland for more than three weeks at one time. One training point that Clark used in training its own technicians was to insist that a new person always take an experienced person along. This procedure was successfully used on the Bumar project.

Host. Clark brought the Polish trainees to its Jackson manufacturing plant where they were given firsthand experience working with the heat treatment operation and gear cutting. The Poles came in groups of ten or less and stayed for three to four weeks. Although a translator was needed, the training was generally on a one-to-one basis. The skill areas represented by the Poles included heat treatment, furnace operation, quality control, and gear cutting.

The number of individuals involved and the length or number of visits were approximately as estimated, Clark reported. Clark did feel that more time was needed for planning and documenting than originally anticipated. In addition, some of the Polish workers wanted training in techniques that were outside the realm of the Clark technology—for example, the manufacture of certain bolts that Clark purchased from others. Thus, Clark could not supply the technology for the bolt manufacture.

Clark found that Bumar lacked skills in application engineering and manufacturing techniques—not in theoretical engineering—and attempted to correct the problem by installing better equipment and upgrading the workmanship. The three-step approach toward improving workmanship involved improving individual worker machining and tooling skills, maintaining tight control over product quality, and attempting to instill worker pride in accomplishment by illustrating how each person's job helps to improve both the product itself and Poland's image throughout the world.[7] These techniques were successful; the product currently produced does meet the specifications set out by Clark.

Clark noted that one problem with the Bumar project was overcapitalization of equipment. Bumar had been suddenly inundated with equipment and required time to learn to use it efficiently. A major step in the right direction had already been taken, however, when the Polish workers identified the problem. Measures were then initiated to overcome these "growing pains."

Other Personnel Involvement

Clark reported no major problems with staffing in Poland. The company's technicians were to be there no longer than three weeks at a time, although individuals could return for additional visits. Clark found, however, that to adequately staff the project, the corporation had to develop a human resources contingent. Then once the project was completed—or, in particular, once the major portion of the training was completed—this group of experienced training technicians had to be absorbed back into Clark's domestic operations (if no other international projects were available). Clark felt that this resource base virtually committed the corporation to seek out additional arrangements that could take advantage of this staff—much as its technology transfers took advantage of corporate research. Clark found that selection of those individuals possessing both the technical and the adaptive skills needed for international operations required skill in itself. The company felt that after these people had gained some practical experience, opportunities should be taken to make use of their services as much as possible.

Protection

Leverage Assessment

As noted earlier, Clark usually did not undertake such an agreement without granting its trademark. This gesture was intended to provide Clark with not only a potential supply source for the manufactured product but also a viable form of leverage in negotiating with its Eastern European partner. This leverage existed because a Western trademark apparently made a product a more desirable export; seemingly, it even heightened the product's demand within the manufacturing country. Whether this general assumption of greater leverage for Western companies was accurate is somewhat in question; perhaps a more accurate assessment would deal with the amount of leverage gained by Clark through the granting of its trademark. Clark technology was certainly respected throughout Eastern Europe, and its reputation surrounding axle technology was excellent; so, in a general sense, its trademark was a desirable item. It might even be questioned whether Clark would have obtained the agreement in the first place if it had not been willing to transfer the trademark along with the technology.

However, unlike complete or final products, the trademark on an axle is far less visible than the logo of a company that appears, for instance, on the front of a dozer or truck. Both Clark and International Harvester transferred their trademarks to Poland for use on their respective products. However, due to the nature of those products and the fact that Clark was involved only with axle manufacturing while IH was producing total vehicles, the Clark trademark was relatively hidden while the IH logo was patently visible. This resulted in more awareness of the IH name than the Clark name throughout Poland—an awareness totally unrelated to product performance or quality. Thus, while a Clark trademark may have an impact on the immediate buyer, the company's general reputation would not be spread as wide or as rapidly via familiarity with its trademark as would be the case with another type of product or the manufacture of an end product.

Clark was not oblivious to this factor, however, and additional measures were taken to enhance the corporation's position. A Clark official described one such enhancement involving the transferred technology. Although the company provides its partner with carefully delineated plans, there is always an undone segment. Even though the plans are delivered piece by piece, a final segment is often missing. In cases of vehicle manufacture, the "missing piece" is often the transmission or drive train technology, which is never transferred. This assures Clark of a continuing relationship because the creation of an adequate transmission usually requires years of study on the part of the other party, and the purchase of the transmission from Clark is a far less expensive solution.

Clark noted that an agreement with Japan was the only one, to date, in which all Clark technology had been transferred. The basis for that concession was Clark's possession of something else that Japan desired—a superb marketing organizational skill. Clark knew that Japan was seeking this knowledge along with access to international marketing entries—both possessed by Clark—as well as Clark's technology. An accompanying factor, as well, was that research in those product lines was rapidly developing new products, or so greatly improving the current ones that they remained desirable and sought after.

Utilized Inspections

An unexpected situation regarding quality control inspections arose in Poland. The Poles were anxious to sell cast-iron housings to the Soviet Union. Clark, strictly adhering to methods and procedures right from the start, insisted on inspections, even though the Poles were not yet manufacturing their own components but were purchasing them from the United States. However, as the items were inspected prior to the sale, differences in quality standards were found among the components purchased from the United States.

Patent/Trademark Defense

Poland cannot sue others for illegal use of the Clark patent or trademark but, instead, must notify Clark who decides the legal disposition. No problem has arisen in this area so far.

Legal Provisions Invoked

An unexpected situation arose with the joint development project in Poland. Poland had a contract to supply the USSR with certain axles. Poland's contract with Clark noted that the transfer of innovations or improvements on the Clark technology supplied to Poland was a two-way street. Both Clark and Bumar were obligated to supply the other partner with any improvements relating to the axle. However, Bumar felt that the improvements made on the axles being designed for the USSR were done totally by the Poles, without any assistance from Clark. Bumar's offer to Clark was strictly limited to the purchase price of the parts of the axle acquired from Clark. At 0.5 percent, this was only one-sixth of the amount Clark felt was correct. The company believed that the improvements made in the axle were possible because of the technology transferred to Poland, and that Clark should be paid at least 3 percent of the price being obtained. This matter was reviewed and renegotiated, and Clark reduced its demand to 1.5 percent but was holding firm at that level. If Bumar rejected this position, Clark intended to withdraw the offer and take the matter to arbitration. The eventual outcome resolved the issue at approximately 1.5 percent of the value, close to Clark's final position.

Hindsight

Factors That Made the Project Easier than Anticipated

Clark felt that a firm should not encounter any major impasses regarding technology transfers if the planning was thorough and concentration was applied to those areas where the license was weak. However, the corporation noted that this planning had to be followed by implementation that was just as thorough, and control had to be maintained throughout the transfer to provide for the interchangeability and quality of the finished products.

The entire operation between Clark and Bumar was described as easier than anticipated because of the good personal relationships that developed between Clark officials and Polish officials of all levels. Clark also found that the Polish employees depended on the Clark trainers for product technology, including instruction; they did not "go off" on their own. This could be interpreted as a reluctance to make decisions, but it provided for a smoother and less troublesome transfer of technology.

Clark found that quality control was easier in Poland than in Romania (but also noted that the Romanian project had not been under way for very long, which would make a difference). However, since this was a pleasant surprise, it made the project in Poland much easier for the Clark technicians.

Factors That Made the Project More Difficult than Anticipated

The two greatest problems Clark faced were "the late delivery of manufacturing equipment to the Polish plants and the slump in axle requirements occurring in Western Europe. This slump in demand, caused by the recession in Western Europe, was passing by 1977, but in its wake left a variety of small orders for various sizes of axles. This type of situation is more difficult for Bumar to fill as the plant is geared toward the manufacture of a higher volume of axles of one type at a time."[8]

One factor that made the technology transfer to Eastern Europe very difficult was the nationalistic tendencies of each nation within Eastern Europe. This prevented the CMEA countries from operating as one monolithic bloc. Each country, given a suitable export commodity, tended to first use the product for its own internal use if it was essential. Then it was exported to the West or to the USSR in exchange for either raw materials or gold. Finally any surplus was shipped to other countries within the Council for Mutual Economic Assistance (CMEA). Clark felt that the corporation would benefit from its axle production in Poland as this would complement the Clark transmission technology being transferred to Romania. However, it was found that interchange between the two countries was not encouraged, and the possibility of exchanges between these countries was relatively minor.

Difficulties did arise from the lack of authority exhibited by the Eastern European decision makers, especially at the local levels. This situation prompted changes over a period of time, however, and the authority to make decisions, especially those not involving any additional expenditure of funds, was provided where needed.

Progress of Future Expectations

It was reported that the reputation of Clark Equipment Company in Poland was apparently not dependent on the visibility of its trademark, for by 1975 Clark was already involved in negotiations in at least six other areas representing possible projects of a relatively large magnitude.[9] However, by 1978, few of these projects had been finalized or awarded to Clark. In fact, one of Clark's major disappointments occurred in late 1975. Additional agreements were concluded between International Harvester and Bumar but not between Clark and Bumar. For Poland, alliances with both companies tended to spread out any risk that might have been present through the use of only one company. Such a variety of associations provided Poland with more resources and contacts than would otherwise have been the case. Bumar's selection of IH instead of Clark for the additional trade was apparently based on a variety of reasons, including several rather intangible factors. One Polish official noted that IH tended to have a more humanistic yet advanced image among the Poles, compared with the image that Clark projected, although no one incident or occurrence could be pointed out as a possible cause for this perception.

Clark's arrangement with Bumar in Poland had an ironic twist. As it turned out, Clark's technology in axles was used in some instances to assemble its competitor's product.

Clark felt that joint projects were quite viable possibilities that were just beginning to be explored. Adjustment to metric scaling was one area in which Polish and U.S. workers could benefit from each other's skills—a conclusion Clark reached as a result of the Bumar project. Joint modification of Polish products for U.S. and Western markets was another resource area, as well as development of new products for both areas.[10] Other cooperative areas might have the Polish partner handling the testing and manufacturing of prototypes while the U.S. partner worked on the tooling and scheduling aspects. Another cooperative venture might be to combine talents in a joint sales effort, using both Polish and U.S. marketing and sales experience for the sale of each other's products.[11]

International Harvester Company

Description Variations

Extensions and Enlargements on the Original Contract

Five years after the contract was signed in September 1972, Jerzy Szumski, economic affairs officer for the Commercial Consulate of Poland, noted that "the feeling on both sides is that mutual expectations have been fulfilled and that perhaps even more importantly in the course of daily implementation of mutual covenants, new perspectives and potentials continually emerge and are not unattended."[12]

The rapport and subsequent developments that materialized during the Bumar-IH project by 1978, ten years after initial negotiations had begun, exhibited several aspects whose very existence spoke highly for this nonequity project. For example, IH has various agreements with several agencies in Poland that are far more complex than the usual license agreement wherein some products are brought in and a royalty is paid back to the parent company, based either on the products brought in or on the sale of those products. Though product importation and royalties were incorporated into the IH-Bumar nonequity agreement, they were only part of the entire operation.

The cooperative agreement has been extended four times to date. The latest extension provided for the manufacture of transmission gears for IH bulldozers that were assembled not only in Poland but also at IH facilities in the United Kingdom and the United States. This manufacturing is being carried out, not at Huta Stalowa Wola directly, but by the Pzlrzeszow Works.[13]

Within three years, the crawler tractor portion of the agreement had been expanded to wheel loaders and pipelayers. As before, this sale of technology was contingent upon Bumar receiving the most modern vehicle information available and the IH trademark.

Additional Products

The progression of products produced in Poland began with the 140-horsepower crawler bulldozers, manufactured in 1973, followed by crawler loaders. By 1975 crawler tractors of increased horsepower, 200 and 285, respectively, were being produced.[14] All in all, IH provided Bumar with seven crawler models.

Also in 1975 a wheel loader and pipelayer agreement was signed between Bumar and IH, adding other vehicles to the production process. The Poles expressed a relatively high demand for pipelayers, a form of crawler tractor used for laying oil or gas pipelines. This brought the number of wheel loader models under this transfer to nine. All were covered by the IH trademark.

This 1975 addition to the basic Bumar-IH agreement was designed to help the manufacturing facilities of both partners. International Harvester wanted to manufacture two units in its British subsidiary: the TD15C tractor dryer and 175C loader. Bumar agreed to provide "skidded tractor units (tractors without engines, controls, attachments, and crawlers) on a multiyear delivery schedule to the IH Doncaster facility for completion and subsequent sale by IHGB throughout the United Kingdom, Europe and Africa."[15]

Joint Development

Product development ranged beyond a simple transfer of knowledge. Mutual design efforts were instigated in May 1975 through the establishment of a joint Polish-U.S. design office for the development of a "new-generation bulldozer and engine." This development contract permits full access to the research and development work of both parties.[16]

Another development project came about because Bumar wanted a different gear drive on the vehicles. Initially, the equipment had a gear drive that could only pull; it could not reverse. The Poles, working with IH, were instrumental in helping to create a power shift version of the machines. Bumar signed a joint development agreement with IH to create a torque converter that would be adapted for use in Poland and other areas with similar terrain. These torque converter components will be manufactured by Bumar after joint development and testing. This project, while a separate entity, is an offshoot of the IH-Bumar agreement. It was signed on March 17, 1978, and is scheduled to run for nearly 20 years, concluding in 1995.

The respective responsibilities between IH and Bumar for the joint development of new crawler tractors, diesel engines, and critical machine components, have been listed as follows: "IH is basically responsible for the engineering design while Bumar is responsible for the construction of engineering prototypes. Bumar engineers will also participate in the designing process as consultants and in order to ensure that the engineering documentation is readily usable in Polish plants and truly metric."[17]

Trademark Permission

In 1977 IH trademark approval was reached for some models of the crawler tractors. The cooperative agreement signed by IH stated that every part and component incorporated in the production cycle, regarding both component deliveries and final products that were to be the property of IH, had to be tested and receive an official conformity certificate. Extremely thorough quality control tests were performed both in Poland and at the IH facilities in Chicago on the first items manufactured within each product area.

IH and Bumar were jointly responsible for developing Polish equivalents for U.S. steel and other materials used in the manufacture of the construction equipment. The Polish facility had a total of 158 steels and metals—of which 147 were found suitable for substitution, while another 7 steels and 4 modular irons had to be especially produced to U.S. specifications.[18] Thereafter, additional checks were performed to make certain that the products were continuing to meet those standards. "It is gratifying to note that the components manufactured through the auspices of Huta Stalowa Wola passed all quality checks and were permitted to carry the IH trademark."[19] The first totally assembled machines permitted to carry the IH trademark produced under Huta Stalowa Wola auspices included the TD15C crawler bulldozers and crawler loaders.[20]

Preparations

New buildings were constructed for the IH project, despite the fact that the Huta Stalowa Wola already was a massive industrial complex. The new fabrication and assembly shop involved 40,000 square meters for the gear shop addition—and more construction under way for the crawler undercarriage shop and a modern automatic forging plant that was being imported from Japan.[21]

In preparation for the IH manufacturing project, Bumar had also created many of its own tools. It was reported that 6,900 items (including fixtures, jogs, tools, gauges, dies, templets, and patterns) had been created by the end of 1973 at an expense of approximately 900,000 toolroom hours.[22]

The technology transferred from IH to Bumar was essentially engineering design documentation, factory standards, voluntary manufacturers' standards, manufacturing engineering documentation for items manufactured by IH, and informal know-how. The engineering design documentation alone consisted of over 20,000 drawings, all of which required conversion to the metric system and translation into Polish. It was reported, however, the Bumar engineers completed this task in several months, building upon former experiences with British-based crane companies—the "soft metric conversion" approach was used to "ensure full interchangeability of parts."[23]

Technology Updates

The IH technology supplied to Bumar has been updated virtually every year since the original transfer. This IH policy of implementing its product improvements internationally on a concurrent basis provides for product consistency as well as interchangeable parts, but it does create some downtime, since all facilities are retooling or making other necessary changes at the same time.

To ensure interchangeability, IH and Bumar developed a coding system to indicate which Polish parts had been approved for substitution with U.S.-made parts. Commendably, 90.5 percent of all possible parts qualified for substitution. This substitution requirement referred to components that were to be used in the manufacturing process, not to items carrying the IH trademark.[24]

Bumar's manufacturing policies regarding the IH technology transfer were summarized as follows:

1. Descend from as assemblies to part numbers levels with imported components when justified by economic and manufacturing considerations.
2. Move from bay assembly to line assembly as fast as possible (a shift from bay assembly approach was in-

troduced by IH to Bumar at the Huta Stalowa Wola facility by 1975 for all five crawler tractor-dozer and loader models).

3. Substitute locally make components for imported ones as fast as technically viable (but only after IH approval).
4. Concentrate on production of substitutes for the most expensive imported items (but ones that are feasible), and gradually build capabilities for the remaining items.
5. Continue to import items that IH also purchases from outside suppliers, or until IH has totally approved substitution.
6. Work toward developing own availability of items not commercially available in Poland that are necessary for product—such as those with peculiar shapes or unique fasteners.[25]

Along this line, it is important to realize that by 1979 some of the vehicles manufactured by Bumar under the IH trademark contained up to 85 percent Polish-made components.

Corporate Structure Revisions

One change that has had an impact on projects in Eastern Europe is the corporate restructuring that took place throughout IH. The previous 34-year-old geographic orientation was based on domestic divisions with an accompanying international division. The new IH structure has a product orientation with five groups that are each free-standing, self-sufficient, worldwide business groups. (The group directly involved with Bumar in Poland was the Pay Line division, responsible for construction and industrial equipment.) Each group contains marketing and technical expertise as well as its own research and development and related service organizations.[26]

Since IH signed the initial agreement, it has "reordered its corporate priorities in the past five years to emphasize overall asset management rather than sales volume." Assets were considered to be virtually anything of value to the firm. This approach involved (1) installation of a new strategic planning program providing strong top-down guidance, (2) implementation of new compensation practices rewarding individual performance, and (3) worldwide reorganization consolidating related assets into product groups.[27] It is this last aspect that directly related to the Polish cooperative agreement, since the Pay Line group included construction equipment.

Commercial Offices

Another result of the IH-Bumar agreement was that by the mid-1970s both IH and Poland had established commercial offices—IH had its Office of Technical Information in Warsaw, and Poland opened the Office of the Polish Commercial Consul in Chicago. The thrust of the IH office was to promote sales of IH products to Poland and to coordinate service capabilities regarding those products. Poland's office provided liaison functions for a number of Polish industries as well as assistance to its representatives. Despite the existence of this office, however, Bumar considered establishing its own company in the United States to help coordinate its activities.

The combine, Huta Stalowa Wola (HSW), was reorganized in 1973 to provide a separate office in Warsaw that would oversee such technology transfers as took place under the IH agreement. This office was created to coordinate such activities as the transfer and absorption of technology, personnel training (for HSW and others from Poland), procurement of financing from Bank Handlowy, acquisition of manufacturing equipment from Polish and foreign sources in cooperation with specialized foreign trade enterprises, and the development of Polish suppliers as sources for various components.[28]

One of the mutually advantageous points of the agreement was that the transfer did not stop with the manufacturing process but also encompassed the area of "worldwide marketing cooperation" for agreement products, utilizing to the fullest extent possible their respective advantages in a marketplace, while fully observing the rights of franchised dealers of both Bumar and IH.[29]

The involvement between IH and Poland has certainly increased since the venture began in 1972. Currently, IH is selling machines made in Poland; that is, IH is using its international marketing capability to export a Polish-made product. This is part of the marketing agreement to support and implement the sales activity of IH products connected with Poland; however, officials note that this involvement has gone far beyond the activities initially envisioned in this area. The company has been investigating markets for its products throughout the COMECON countries and elsewhere. In return, for example, the company is marketing Polish products in Africa. Currently, construction machines produced through this cooperative agreement have been purchased by users from Yugoslavia, Libya, Angola, Algeria, and the People's Republic of China—in addition to sales to the Soviet Union. These sales are in addition to those construction machines marketed internationally through the IH distribution network.[30]

Conclusion

The joint development projects that grew from the IH-Bumar agreement are an indication of how much broader the project has become beyond what was originally anticipated. By 1977 IH officials reported that the bugs had been worked out and that things were going quite well on the Polish project. This was a change from earlier, more pessimistic reports. As projected, subassemblies from Poland were being shipped to Great Britain for use in the IH facility there. Thus, even disregarding the technology fee as a source of profit, IH felt that the corporation had benefited more than expected. It was selling goods that it otherwise would not have sold. It had developed European sources for IH products that IH could not have afforded to do on its own because of the great amount of capital required for such a venture. Furthermore, the joint development projects had expanded the scope of the initial one-way technology transfer into a two-way international information flow and research partnership.

Current Corporate Perception of Benefits from Agreement

IH's initial motivation for entering into the Bumar agreement was the financial gain to be made. Hayden had reported that IH had three aims in this regard: to sell entire vehicles until the Bumar project was onstream; to supplement that production, once it was going, through the sale of components that were not locally manufactured but still necessary for the vehicles; and to continue to sell replacement parts until an adequate supply of service parts was in stock. Hayden stated that Bumar was obliged to purchase $1.5 million worth of components in the first three years, followed by another $6 million for additional components as agreed upon in 1974.[31] These anticipated financial aims ranked first, followed by the profit gained from the technical disclosure fee and by Bumar's potential to become an additional source of products to supplement other IH manufacturing facilities.

Now, seven years later, IH still ranks the sale of products as its number one priority—the significance of this is explained by the fact that, until the initial Bumar cooperation agreement in 1972, IH had no market entry or access to the socialist bloc. However, even the sale of products is losing its number one position to product source availability. After an initial period during the setup and training time, which involved some problems and confrontations, IH has become increasingly complimentary of the skill and workmanship of the Poles. Bumar's ability to adequately manufacture vehicles according to IH specifications has provided IH with a dependable product source. In 1976 one IH official termed Bumar an "absolutely first-class source which is going to be one of the main manufacturers of crawler components, replacement parts, and final products for all our markets, especially Western Europe."[32]

International Harvester admitted that, in general, it did not glimpse all the long-run possibilities of its agreement with Bumar. The agreement was merely taken to be a type of license arrangement, even though the technology being transferred was far more extensive than in usual license agreements. Only a few executives appeared to grasp some of the greater benefits that might be gained through the association.

The market for replacement parts and service for construction equipment was apparently an area that could be developed more fully. The socialist bloc appeared to have a dearth of replacement components because most parts were manufactured for new products; IH felt that the service side ought to be expanded. Sales of secondary parts had not been a major consideration in the earlier phase of the agreement, but IH felt that this aspect would become increasingly important as more machines were used throughout the Eastern bloc.

Negotiations

Contract Format Revisions

Negotiating for Huta Stalowa Wola was the Polish Foreign Trade Enterprise Bumar; the contract with IH was one of the first industrial cooperation agreements with a foreign partner ever signed by the Polish industry.[33] It was noted that "the initial talks were accompanied by prejudices resulting from ignorance of one another's possibilities." However, it was found that reserve and tension disappeared within a few years when it became clear that both partners had proved themselves worthy of each other and "mutually indispensable."[34]

The term indispensable may not be as accurate as mutually advantageous, since both parties benefited from the union. IH established a bridgehead on the vast market of socialist countries without having to develop its own manufacturing facilities there, and its Polish partners technically improved their construction machines and were able to manufacture products sought by buyers from hard-currency markets.[35]

The initial contract drawn up between IH and Bumar was the result of long and difficult negotiating sessions. IH reported that terms kept changing, amendments were added, and fees had to be recalculated accordingly. However, since that opening experience, IH found subsequent contracts or revisions for extensions or additions far easier to formulate. The contract format was changed and is now considered much easier to follow and understand. Before, the wording and intentions were expressed in general terms; now, the agreements are extremely specific in all regards, including any marketing obligations. IH noted, however, that while these more recent versions might be more suitable, its past experiences were necessary for this evolution to take place.

IH found that quality standards were difficult to enforce: daily inspections had become commonplace in order to meet quality specifications. The company stated that all of its future agreements would delineate inspection and quality standards; it had learned that all details had to be accounted for in order to guarantee quality.

International Harvester emphasized the fact that in a nonequity transfer of this magnitude, it was important to understand the people with whom one dealt as well as the products being transferred. The corporation noted that in arrangements with most countries, but particularly areas such as Eastern Europe, the mutual benefits of any arrangement must be emphasized. A viable contract must contain advantages for both parties, otherwise one partner will become dissatisfied or disgruntled with the agreement. Resources should also be listed, including people available. Leaders of the country must be involved, informally as well as formally, during negotiations. IH stressed the fact that Poland's politicians were involved whenever possible.

One of IH's first indications that this project with Bumar might be more time-consuming than initially anticipated emerged in a negotiating session that took more time than expected and involved more people. The original negotiations seemed to involve a small number of people, but as the negotiations intensified, an expanding, inverted pyramid of individuals were contacted as various operations were considered. These individuals included engineers and financiers, as well as the manufacturing/technical staff.

Current world events were on Poland's side in the early 1970s with regard to this agreement. A worldwide depression of sorts had affected the truck and heavy equipment business when the Polish came to IH. International Harvester had reached a decision point; by the early 1970s the company "severely needed capital and could find virtually no sources."[36] In an environment of depressed market activity, IH was persuaded to take the risk of providing a potential competitor with not only current but also future technology relating to its construction equipment line—in return for certain advantages offered by Poland.

Technology—including the latest improvements—was a firm requirement of the Poles. Bumar's Managing Director Zaskurski remarked that throughout all Bumar negotiations and agreements, "one basic principle has been always adhered to; namely, that any product to be manufactured within the Bumar Union either as a final product or an assembly had to be a product to be yet manufactured by the given firm in the future and not a product the production of which was soon to be abandoned." He also noted that "all cooperation agreements have been concluded only at the stage when the given product was either still on the designers' drawing boards or just implemented into production."[37]

Corporate pressure was evident throughout the negotiations as Poland insisted upon current, virtually future, technology. The fee for that technology was high: IH officials said $100 million for the company's technology; Poland offered $10 million. The fee finally agreed upon was a compromise between those two figures.

There were also business pressures on IH. The changing business scene, even in the area of pollution controls, was a factor as IH had assumed it would have 10 to 12 years to incorporate controls measures. The company then found it had three to four years instead. Thus, certain technology would have an even shorter life span.

International Harvester pointed out that while the wording of the contracts and extension clauses became more specific with regard to duties and details, the overall extent and intent of its agreements with Poland broadened considerably from its initial strict, finite agreement to produce crawler tractors. Now IH states that the corporation sees its agreement with Bumar, via their joint development projects, as cooperation on an entirely new plateau.

Pending Negotiations

No new negotiations were pending, according to IH officials.

Time Frame

On Schedule

It is important to note that all payments from Poland, due on fixed dates, were fully completed on time. The various projects undertaken by IH in Poland, however, met with several delays. Since a specific time frame was not incorporated into the agreement, these delays were distinguishable only in contrast to what was generally expected, not to certain deadlines. The emphasis was placed on the achievement of goals. Because of the "end result orientation" both Bumar and IH reported that they did whatever was required to achieve those results. In some instances, those goals required more time than initially estimated, but it was felt that it would be inaccurate to view this as a negative factor.

Behind Schedule

The agreements did not contain many time deadlines, according to IH officials, but estimations were given as to the likely period of time required for the operations. According to this estimate, the setup time for the crawler tractor production ran about two years behind schedule, using finalization of manufacturing capability as the end point of the setup. Part of the reason for the extended time was apparently due to the lack of an infrastructure for implementing decisions. Consequently, many workers and supervisors hesitated, afraid of making the wrong decision, which tended to detract from general efficiency.

The anticipated time frame for the wheel loaders was less accurate than for the crawler tractors. This apparently was due to the fact that the entire line of loaders was redesigned, a process that benefited the ultimate product line but that greatly hampered the schedule as projected for Poland. The design holdups were caused by IH, not Poland, and IH took full responsibility for any delays connected with this aspect of the transfer.

Time schedule expectations also varied between IH and Bumar. At one point, Bumar was 30 percent behind in terms of its crawler tractor production and felt that overall production was two and a half years behind schedule. However, IH hastened to point out that Bumar's projected schedule was not necessarily the schedule anticipated by IH, as the corporation suspected production would take longer to get on-stream.

One problem that is likely to arise in any instance when production is dependent on certain component imports presented itself in the Polish project. Poland's cutbacks of Western goods, including components essential for the manufacture of the IH wheel loaders, slowed production of the vehicles. Indirectly, this slowdown also affected the expansion of other facilities at Bumar and thus had an impact on all of the production and assembly connected with those components or projected expansion plans. However, this type of balance of payments and trade problem, while outside the total control of those involved with the project, is often of a fairly temporary nature. Its impact would be equivalent to the effects of an economic slowdown.[38]

Other Aspects—Variations

Variations occurred in the time laid out for the agreements. The crawler tractor contract was signed in 1972 and was to run for a period of ten years. However, the additional units—including wheel loaders, which were added in September 1975 via an amendment to the crawler tractor agreement—provided a concurrent term that was to be not less than 15 years. Thus, the initial 1972 agreement was lengthened and its termination date extended from 1982 to 1990, along with the wheel loader amendment. The joint development projects were scheduled to extend until 1995, five years beyond the completion date of the other two projects.

Financial Flows

Payments

Payments from Poland to IH have been exactly on time as designated. The only payments that were variable were related to production schedules, such as royalty payments involving purchases of components used in the assembly of the final vehicles, or IH's repurchase of certain components manufactured in Poland for use in IH production.

Payments have been in U.S. currency, although the English subsidiary of IH initially used Swiss francs. No payment problems have occurred as yet, but it was pointed out that, in agreements of this nature, there are three general sources of revenue. The first is the disclosure fee or initial payments. The second is profits from sales of products or component parts, usually most plentiful before the facility is on-stream. The third source comes from royalties from the sale of completed products or components. This last area carries with it the greatest risk because the receiver of those royalties maintains the least amount of control over the situation, and it also involves the longest time period. The longer the time period between the contract's signing and the receipt of the fees, the more risk is incurred.

The earliest years of the Bumar project were the most severe for IH in terms of revisions to the corporation's original expectations. The disclosure fee was not all profit, free and clear. It was noted by the mid-1970s that "the expense of travel back and forth, the need to hold the Poles' hand every step of the way, the executive negotiating time, and the mammoth training effort required, had all but erased any expected profit from the technology fee."[39] However, since that time, the corporation has reevaluated the expenses involved in training and negotiating and considered them to be more of an investment—fruitful investment, because a product source had been created. Furthermore, a portion of the technology fee did remain after all training had been accomplished on the Bumar project.

IH also encountered greater expenses in equipment testing, particularly with regard to some axles that were sent to the United States for additional strength tests. This was in addition to the higher costs associated with the additional travel to Poland.

IH stated that the corporation was, in general, very pleased with the Polish project; but the anticipated volume, both in terms of production and in terms of sales, was not as high as earlier hoped. This was true of IH's expectations, but also particularly true in terms of Polish expectations. It was noted that while the Poles had talked of thousands, they were producing hundreds instead. The demand, however, had not been heavy either. Product demand was expected to pick up, however, since worldwide construction was steadily increasing.

Sales of Bumar-produced equipment throughout Eastern Europe were arranged through Polish sales offices. IH was to receive a royalty from these sales. The corporation also received some publicity from these sales. The IH image expanded as familiarity with the company grew throughout the region. In fact, the importance of IH's Eastern European dealings had been reflected in its corporate structure, which had taken increasing note of the agreement with Poland by placing additional importance on sales to socialist nations.

Purchases from the Domestic Company

While it was true that the anticipated profits from the agreement had been offset by the extra man-hours required, the products being developed in Poland were expected to result in additional IH profits. The equipment produced under the Bumar agreement at first contained between 20 and 30 percent Polish content; thus, IH was contributing 80 percent of the components in some of those vehicles. However, by late 1978 some models were made of 85 percent Bumar-made components.

IH sales directly to Eastern Europe, as a result of the Polish agreement, were limited to a few replacement parts, although some small crawler tractors were purchased by Poland from the English subsidiary during the negotiation period from 1968 through 1972. This was done more as an example of the good faith being generated by the talks, however, than as an indication of future sales through this channel. Even when all manufacturing was on-stream, Poland was still slated to buy some components from IH, since the anticipated Polish national content of all machines was to be 60 to 70 percent, with some of the remaining components continually purchased from IH.

Poland's financial flows were apparently further below expectations than IH's, since Bumar's sales were related to the production schedule. With anticipated production running a year to 18 months behind, Polish revenues from IH buy-back purchases were also behind. However, this was an area that would change as production picked up. It appeared that the Polish approach toward delivery of some products was "if not today—tomorrow"—whereas, in keeping with certain schedules, the promised orders had to be received on time.

Purchases by the Domestic Company

International Harvester did purchase a number of items to help its employees who were sent to Poland. Those purchases included houses (several villas) and IH's line of consumer vehicles, International Scouts. In addition, IH also financed extensive rest and recreation holidays for its employees in Poland. Certain foods and other consumer products not available in Poland were shipped in by the company as well.

Personnel

Training Assessment

Domestic Company. IH's staffing requirements varied from its initial expectations as the Bumar project progressed—greater involvement was required, utilizing more people from IH and a greater number of trips to Poland. The length of the visits remained fairly constant, however, with regard to expectations, as did the job areas of the personnel being utilized.

IH's current arrangement in the United States has been to assign three or four individuals to the Polish project, drawing upon others there at IH as needed. This was in addition to the IH representatives placed in Poland for a period of time. The company found that these overseas placements detracted from the technology transfer profits, but they had the overriding priority of making the project successful. For the Bumar facility to be ultimately profitable as a product source, IH realized that it might be necessary to forego some immediate gain to achieve the long-term goals.

Throughout the technology transfer from IH to Bumar, only two IH resident personnel managers were assigned to Poland on a lengthy basis. Poland viewed this as "something generally unheard of in cases of a technical venture of such complexity. It was a good testimony to the high levels of technical proficiency of Bumar's plant and its people."[40] By 1979 the IH representatives in Poland consisted of a territory manager and a service manager.

As anticipated, International Harvester did encounter some resistance from some of its employees toward being sent to Poland. However, the corporation worked to provide a decent environment for its employees sent there. No additional salary or financial fringe benefits were supplied to these employees, but IH provided the additional living comforts as described above. These benefits helped those employees save a greater portion of their salaries than would have been generally possible in the United States.

The selection of corporate employees to be involved on an overseas project is a careful consideration at IH. No specific training courses are used, as IH apparently feels that "adjustment" seminars are "almost useless," according to some IH officials. Instead the corporation prescreens possible employees for overseas positions and uses personal counseling involving the entire family. However, the increasing expenses (including difficult taxation payments) associated with sending employees overseas for any length of time have finally forced IH to severely cut back on such assignments—to the point of virtually discontinuing sending U.S. citizens overseas. Monetary expense is not the only justification for this cutback, as it has been found that the comfortable environment in the United States makes it especially difficult for many of its citizens to adapt to conditions abroad. Other nationalities, such as the British, Germans, Australians, or New Zealanders seem to fare better.

Previously, the presence of a host country national as the overall manager of an IH operation abroad was an exception; today it is closer to the norm, as IH actively seeks out nationals for such training or helps persons from another country apart from the United States adapt to the position. Americans are often sent abroad for relatively brief, fixed time periods instead.

Host. Purchasing a license from the U.S. company required a number of changes and adaptions on the part of the Poles. This included conformation to U.S. documentation and specifications and adherence to tests to verify high product quality. It also required the modernization of production machinery and the purchase of special-purpose machines, including numerically controlled machine tools designed to ensure high precision of individual components.[41] Heat treatment and welding techniques were modernized, and personnel were updated as to the use of the equipment. On-the-spot training and inspections were provided by visiting specialists from International Harvester.

IH had stated that "never again will it enter into a technology transfer agreement based on expectation of penetrating a new market . . . (but rather that it would only consider such agreements where it had advance assurance that its partner would become a first-rate source of the product)."[42] However, since 1975 (when this thinking reflected the outlook of IH) the corporation apparently gained new confidence in its ability to transfer techniques and skills and expressed additional interest in other similar ventures throughout the Eastern bloc. Poland, through the actions of the Bumar employees, had been partially responsible for this increase in confidence on the part of IH, because the later outcomes of the Bumar project were proving very successful.

IH found it was not prepared to meet the extensive training needs of Bumar employees. These needs did not stop with the top managers but extended down through all supervisory levels. Programs were needed to build enthusiasm as well as skills and quality control measures.

IH found three types of training difficulties present in the Polish project, despite Bumar's prior experience with Western companies. Poor quality control procedures were the major difficulty, but this was aggravated by a low level of skills on the part of workers and also by inappropriate worker incentives. IH found itself instructing workers not only in the manufacturing process itself but also in managerial or supervisory skills, especially as these related to quality control procedures.

The initial documentation fee was increased so that IH could provide the Poles receiving training in the United States with a per diem allowance for living expenses. The Poles were starting from a far lower level of expertise than originally anticipated, but it was found that the anticipated number of man-hours was more than ample for transferring the technology. For example, it was found that the Poles had no actual concept of the type of building required for this kind of operation—for example, requirements for the floor strength or sanitation, including shower rooms. IH had expected this knowledge to be part of Bumar's basic knowledge.

International Harvester found that the Poles were not used to working up to the standards required by IH; therefore, training measures went beyond assembly line techniques to such points as teaching the Poles to cover equipment at the end of the shift or similar precautionary measures. Part of the problem was that the Poles were not used to having to adhere to such strict tolerance levels and previously had not needed to follow procedures so exactly.

Qualtiy control standards did not relate only to materials tolerances and the workmanship of the final product. In many manufacturing operations, cleanliness affects the output and such factors as the dust in the air may be crucial.

One aspect of training that frustrated IH officials was that the Poles, especially at first, did not follow instructions or ask for advice. One official noted that even though the Poles had purchased IH technology, they did not act as though they actually wanted that information. Instead, they kept "reinventing the wheel."

On the other extreme, the questions asked by the Poles in the training sessions went far beyond the actual need-to-know areas. For example, they wanted to know all details of a bolt, including the chemistry of the steel. This type of questioning would be excellent if the project were to be manufactured, and such questions would also apply if the product were to be custom ordered, as certain specifications might be required. However, when the items are basic purchases from third-party suppliers, such questions may be more distracting and confusing than beneficial.

Another area that required more attention than originally anticipated was that of tooling procedures. The Polish workers had a high familiarity with the operation of some of the equipment, but tooling up for the new products was an area that required additional training.

Troubleshooting also needed more attention than originally anticipated. Trainers from IH spent a great portion of their time checking to see where manufacturing problems were originating from and determining the cause behind malfunctions or parts that were not coming out properly. However, these problems were often temporary ones, solvable when correct operating procedures were explained and then followed. Part of the training involved untraining—prior procedures different from IH methods had to be sorted out and disregarded before the new methods were applied. IH had not anticipated this revised training prior to the agreement, but it became necessary as the project advanced.

In general, International Harvester found the training sessions far more involved than originally anticipated. The Poles had their own management style and techniques, and IH trainers had to incorporate some of that style and those techniques into the new material. The training ranged far beyond the usual product-process information transfer. IH documentation had to be interpreted, substitute materials had to be selected, and all know-how related to processing the necessary parts had to be relayed.

Some of the training difficulties appeared to originate with the observation that "few planners were evident among many plodders," as one person described the situation. Supervision appeared limited to giving orders and such orders were usually followed, often regardless of the circumstances or in the face of apparent problems.

Other Personnel Involvement

International Harvester was responsible for helping the Poles merchandise the product as well as produce it. However, merchandising in a socialist sphere is unlike operations in a more free-enterprise economy. Many sales were from government to government or agency to agency and were often not based on a profit motive. However, cash flow was very important, based on the Eastern European need for hard currency. One effect of this was that the Poles needed additional training in adequate equipment comparison because they lacked the advantages of product knowledge gained by competitive selling. Thus, IH also became involved in teaching comparative shopping for products.

IH hoped to hire some of the Poles as salesmen for the construction equipment to work with potential customers in Poland and elsewhere in Eastern Europe. The strategy was to familiarize possible users with IH equipment so that it would be requested by customers, not forced upon users who might want a competitor's product instead. This "pull" strategy was obviously built on IH's worldwide reputation for excellence, and it helped to further this image. However, by training the Poles in sales techniques, IH was also transferring information that was secondary to the product technology itself—even though it impacted on the technology transfer by encouraging the sale of the products.

Protection

Leverage Assessment

Domestic Company. International Harvester maintained a high degree of leverage throughout the entire project with Poland. This was due to several reasons. First was the fact that all training, in general, took place in the United States, with only temporary visits to Poland by IH specialists. Therefore, delays in training and production affected Poland far more than IH (in terms of personnel, production revenues, and other financial flows). This is not to say that IH was not affected, since the company certainly noted a far greater involvement in its training responsibilities and personnel than originally anticipated. However, Poland had more to lose through major delays or holdups.

Great teamwork resulted from both parties doing their best to instigate production as rapidly as possible. This teamwork brought about a greater rapport, which, in turn, grew through the multiple joint development projects that evolved during the first few years of the agreement. Such rapport may seem the opposite of control or leverage, but in reality, the mutual desires of both parties to work together might be viewed as leverage directed toward getting the job done while maintaining a high self-image.

The IH trademark apparently afforded a high degree of control in this instance, since the Poles greatly desired the use of the trademark on vehicles completed at their plant. This trademark was seen as beneficial to the Poles in terms of their increased export market. It was virtually necessary for IH, since the facility's production was to be a major source for both some assembled vehicles and component parts. Should the products not be manufactured according to specifications, or should other problems develop on the project, the potential loss of the trademark would be a severe detriment.

Required component purchases from IH also helped to maintain an aspect of leverage in that Bumar required certain components to completely assemble the vehicles. Inability to obtain these components from IH would jeopardize the entire project's production capability for total vehicle assembly.

Total vehicle buy-back was a form of leverage, too, in that IH had the option of refusing any products that did not conform to specifications. Thus, IH had some control over the quality of the products, in addition to lending its trademark, and since Poland's desire to sell the products was high, the buy-back option was also important to the financial status of the Polish operation.

Variations in IH guarantees were due to the increasing amount of technology being transferred as more vehicle and component information was involved with subsequent amendments. IH had offered some of the research capabilities of the corporation with regard to the joint development project, in which all information was to be shared between the two partners. IH had, for all general purposes, completed its initial guarantees to transfer the technology necessary to manufacture crawler tractors and wheel loaders when the vehicles were permitted to carry the IH trademark.

Host. As it developed over the first seven years, several aspects of the project helped to identify Poland's control factors—in particular, Bumar's—as well as IH's. One factor was the teamwork issue: whether the partners were with the joint development project or with the assembled vehicles that were to be purchased back by IH for use in meeting its international marketing demands, IH and Poland would both benefit from their mutual efforts.

Poland has the potential to be a true competitor of IH, but the issue has been, at least for the immediate future, postponed. The IH trademark and Poland's current dependency on IH for engines and other components reduce the competitive angle.

Actually, Poland's greatest control measure is a result of IH's increasing emphasis on the facility as a growing source of products to meet its marketing demands. The more dependent IH becomes on the Bumar facility's output, the more leverage Poland gains.

Poland's guarantees, through Bumar, were met with regard to facilities, equipment, and trainees necessary to manufacture the desired construction equipment and vehicles. Again, as above, Poland's expanded guarantees related to research capabilities involved in the joint development project.

Utilized Inspections

Because the construction equipment had to be manufactured to withstand great amounts of stress, and because the IH trademark was involved, inspections and tests were a major concern throughout the project. Initial tests on the components and on the final equipment—as well as on the raw materials used in the construction of the equipment—were extensive, both in Poland and in the United States. Since some vehicles have now been certified as meeting all specifications and standards required to carry the IH trademark, tests and inspections will be limited to maintenance of those standards and to certification of additional vehicles or models as available.

Trademark approval was permitted only after the Polish-made vehicles had passed extensive tests conducted at the IH testing grounds outside of Phoenix, Arizona. As of 1978 three vehicles (TD15C, 175C, and TD25C) had been approved for the IH trademark.[43] Permission to use this trademark was noted by IH Board Chairman Brooks McCormick at the celebration of the fifth anniversary of the IH-Bumar project: "We never would approve its use by anyone if we felt it would not continue to be respected. Bumar and the employees at Stalowa Wola have earned the right to use this trademark through their outstanding performance over the past five years. . . . It represents the highest recognition we can bestow upon one of our trading partners."[44]

Patent/Trademark Defense

Bumar used its own trademark when selling its construction machinery in joint markets (presumably those markets in which it held a nonexclusive license to sell such machines), but those tractors did not carry the IH trademark. The two trademarks could not be used together on the same machine.

One IH official noted that the Poles covet the U.S. trademark far more highly than most Americans realize, and far more than expected even by IH.

Legal Provisions Invoked

No situations requiring arbitration or other legal resource had arisen.

Hindsight

Factors That Made the Project Easier than Anticipated

International Harvester found that the flexibility and dedication of the Polish people certainly made the project not only easier, but probably feasible. The corporation felt that this overwhelming desire on both sides, from the highest level on down, to have this transfer work successfully was the ultimate reason for its success.

The corporation found that this dedication transcended the difficulties that were encountered, whether in the training area, equipment operation, or even in the wording of the contract. Wherever certain portions of the contract were found to be vague or misleading, the individuals involved were determined to identify the end results to be achieved, and work proceeded from that position.

Factors That Made the Project More Difficult than Anticipated

From the viewpoint of IH individuals closely involved with the entire operation, perhaps the most difficult aspect of the project was getting the Polish side to accept the quality standards required by IH and to apply proper control policies and procedures. It has been pointed out that IH was in a beneficial position with Poland in that a British firm, Jones Cranes, Ltd., had previously entered into a technical assistance agreement with Bumar and had provided an earlier introduction to Western-type demands. Jones had agreed to accept the completed product as partial payment, so the Poles had been held to strict reinforcement of the British firm's quality standards, virtually Bumar's first contact with Western quality control.[45] The rejection rate had been relatively high, but the lessons learned and the skills developed carried over into the agreement with IH.

From another angle, however, the prior training given to some Bumar employees was a deficit, because the employees automatically did things one way when IH required a different procedure. Thus, much of the training involved unlearning, helping the employees to ask questions or to seek instructions instead of operating on their own.

The nonequity aspect of the arrangement made it additionally difficult for IH in a training sense. Since IH had to achieve its specified quality standards through the efforts of the Bumar workers, it could not just send in its own team of specialists to correct any deficits that might occur. However, despite such frustrations, in the long run this aspect has proved to be a great advantage. The Poles have mastered IH's quality control techniques and are able to produce products meeting all specifications. Thus, IH now has a supply source that operates on its own, without any equity commitment from IH.

Progress of Future Expectations

One source summarized IH's benefits from the IH-Bumar project as follows:

1. A significant increase of sales of construction equipment to countries of Eastern Europe, including the USSR (1,000 heavy crawler tractors in the past five years), Poland, Hungary, and Yugoslavia;

2. An excellent image among countries in the Eastern bloc, where a similar cooperative agreement has already been established with Hungary, and more under negotiation;

3. Additional revenues from license fees, and commissions on some Bumar sales;

4. Expanded market penetration through creation of new sources for its products in Poland and the United Kingdom;

5. A stable and growing business of selling components to Bumar (although Bumar's shopping list is becoming shorter, the annual values grow with the increasing production of the Polish plant);

6. Creation of an economically attractive source of components and assemblies of guaranteed quality;

7. The ability to introduce some new products and own-design components at one-half or less of normal product development cost.[46]

Advantages to Poland and Bumar from the IH agreement have been described as follows:

1. Important gains in manufacturing and engineering technology;

2. An influx of valuable knowledge in the areas of product marketing and after-sales service;

3. Improvement of professional skills of engineers, workers, sales, and management people;

4. Availability of first-class products with vast potential of sales in Poland, Eastern Europe, and other countries;

5. Valuable and growing opportunities for supplying components and subassemblies to IH plants in the United States and elsewhere;

6. Opportunities for deploying own engineering and product development resources for creating jointly with IH (highly) competitive and advanced new products with worldwide acceptance, at a substantially lower cost and risk than in the case of its own development.[47]

Since the venture with Poland turned out to be beneficial far beyond initial expectations, IH has revised its expectations for the future. For a start, IH has begun to see the advantages of the non-equity type of technology transfer, particularly when the craftsmanship and determination of the host party are as strong as that found in dedicated Poles. Because of their skill, equipment is now being produced in Poland and purchased by IH as an integral part of the corporation's international marketing mix.[48] It has also been reported that, based on the success of manufacturing components for the IH facility in England, IH is now exploring the possibility of manufacturing components in Poland for use in its manufacturing facilities in the United States.[49] Hayden noted in 1976 that IH would "enter another technology transfer agreement if assured that their partner would eventually be a first-rate source of supply of the product."[50] This opinion has expanded since that time; IH now feels confident of its ability to transfer the necessary training and skills to the point where nonequity agreement situations are indirectly being sought out. Yugoslavia and Romania are examples of two countries where interest has been expressed in possible technology transfers. International Harvester has expressed a growing interest in both Yugoslavia and Romania and has increased its product line offerings in an effort to fulfill projected needs in those countries. It would be accurate, then, to point out that IH is now considering additional technology transfers within the socialist bloc, and it considers the nonequity type of transfer a viable means toward accomplishing this end.

Honeywell

Description Variations

Honeywell's experiences with Poland thus far have been excellent. The agreement between Poland and Honeywell, Limited, the Scotland-based subsidiary, has been extended for another year, and overtures have been made to Honeywell headquarters in the United States for an extension of the 1974 agreement. Regardless of the length of the requested extension of this technical assistance agreement, the total transfer of the technology—including the valuable sensor information—should be completed as of December 1979.

Since no intervening events occurred to disrupt relations between Poland and the United States, the security measure of the sensor information in the bank vault in Switzerland will not be a necessary aspect of this final portion of the transfer. Instead, the information in Switzerland will be sent back to the United States; that information dates back to 1974, and updated information, as of December 1978,* will be assembled and sent to the Polish representatives in early 1979.[51]

The initial operation in Poland went extremely smoothly, according to Honeywell personnel involved with the project. Part of

---

*An earlier analysis of this Honeywell-Metronex agreement by Hayden (Technology Transfer to East Europe: U.S. Corporate Experience [New York: Praeger, 1976], p. 60) noted that the sensor technology transferred would date back to the 1974 state of the art. This is not so, as Honeywell is providing Poland with current technology.

the credit for this smooth operation is attributed to strict adherence to the agreement, including all specifications and procedures. One of the few variations that Honeywell permitted was to allow Poland to purchase some of the equipment required for the project in Poland (or elsewhere in Eastern Europe, if it could be located there), instead of ordering all equipment from Honeywell in the United States. This variation aided the hard currency flow from Poland. There apparently has been no detrimental effect in terms of equipment quality, nor was the timing aspect affected. (It may even have been helped by the relatively local purchase origins.) Honeywell did have to approve all the purchases.

Not only did Honeywell require strict adherence to the terms of the agreement, it also required that all changes related to updating the product be made as soon as changes were able to be documented in the United States. This was required, even if the application of any changes required additional training in Poland or even in the United States. To avoid mistakes, the Polish engineers were required to use Honeywell drawings.

There were no changes in any of the major elements in the agreement. The agreement has gone very smoothly and has been more profitable than anticipated, indicating that the company will probably encourage an extension of this agreement.

Current Corporate Perception of Benefits from Agreement

The profitability of this venture has surprised Honeywell beyond the company's expectations. The sale of products has been particularly beneficial. During the negotiations it became evident that while Honeywell preferred the use of a pricing formula throughout the agreement to adjust the price of the units to the current worldwide economic situation, the Poles desired a firm price for each item, to be listed in the contract and constant during the agreement. Honeywell acceded to this request and supplied a constant price list. As it happened, the price of the unit fell below this price (an unusual occurrence in these times of inflation), but both parties had to stand by the stated price. Furthermore, the price quoted for the subassembly pieces was the same as for the completed assemblies, but the Poles preferred these subunits for use in their training sessions. It was reported that "25 percent of Honeywell's price is based on assembly costs,"[52] so this turned out to be a windfall (although this was not confirmed by the company). Thus, the fees gained from the sale of its products, combined with the fees paid for the technology, have made this transfer particularly successful.

The success of this venture has encouraged the company to extend consideration of the current agreement as well as to look for additional ventures in Eastern Europe.

Negotiations

Contract Format Revisions

The success of the agreement has been partly attributed to the efforts of the negotiating team. The one general criticism that Honeywell had with the Metronex agreement was that the conclusion of the contract was not worded as crisply or precisely as the initial portions. However, since this portion dealt with, or referred to, procedures involved in the potential extension currently being negotiated, it was hoped that any ambiguities would be cleared up during those discussions.

Honeywell felt that the formality that existed and was followed at the start of the negotiations was beneficial, in that the stipulation to follow all contract obligations precisely was thereafter adhered to rigidly by the Poles. The protocols for the agreement were completed on the spot, and only minor revisions have been required throughout the working of the project. An example of such a minor variation would be the changing of the name of the Polish representative in the United States for this project.

Pending Negotiations

The current agreement has a high probability of being extended, and other agreements with several other countries will be under way shortly.

Time Frame

On Schedule

The time frame for the 1974 agreement was carefully adhered to in most respects.

Behind Schedule

There were only minor timing variations, within two to three weeks either way, regarding the setup time framework. One instance occurred when a minor problem during the construction and equipping of the Polish factory threatened to delay the entire project. However, the Poles were determined to keep the project on schedule and put in extra hours. To do so, Metronex actually ran three shifts for weeks to get up to schedule. Because of this additional effort, the plant's construction was completed in record time, which kept the project functioning according to the projected schedule. This was essential, since the Honeywell training representatives would not come over until all was ready to go. This devotion to the project on the part of the Poles provides an insight into the success of this project. All involved with the agreement wanted it to succeed. Thus, extra effort such as working overtime to keep within the time frame was done on both sides to help ensure its success.

Another problem occurred with regard to a production schedule, which had its origin in the product's design. It is important to note here that it is most uncommon for a company to agree to transfer its most recent or current technology. Part of the reason for this reluctance is, obviously, to take advantage of its own research and development before permitting others to use it. Another reason is that the transfer of technology is itself a challenge, and it is preferable to have all aspects of the product perfected before such a transfer is made. In this instance, the problem occurred when the development of one of the variations of the class 41 transmitter incurred some problems, causing the timing of this product's schedule to slip. This eventually resulted in delays while the product was being revised under Honeywell's auspices in the United States. Thus, it might be said that birth pangs were still coming after the product had already been sold. However, Honeywell worked to overcome the problem, putting in extra effort and hours as the Poles had done, and these delays were surmounted.

The Poles' demand for the very latest technology has proven to be to their benefit. Now that the initial agreement has reached its conclusion, Metronex has a piece of technology that is relatively unique and as up-to-date as possible. Honeywell's willingness to provide this technology and work with the Poles through any difficulties also provides an indication of the concern that the corporation applied to the project.

Other Aspects—Extensions

The extension requested for the 1974 agreement is for another five years and includes a request for updated information on the class 41 transmitter during that period. However, the updating through this additional period will not extend to sensor technology. The sensor is still a unique product and highly viable. The Poles, through their insistence for the most advanced and latest technology, have obtained a valuable product.

Despite obtaining the sensor technology, the decision to actually manufacture it has yet to be resolved by Metronex. The cost of tooling and training would make the manufacture of the sensor itself an extremely unprofitable undertaking, at least in the immediate future. Thus, despite the transfer of the sensor technology, another request (accompanying the request to extend the agreement) has been proposed for a ten-year purchase agreement (to 1989) of the sensor assembly. Currently, there is a ceiling on the number of sensor assemblies that may be purchased each year, set at 2,000 sensor assemblies per calendar year.

Financial Flows

Payments

Financially, Honeywell has found that the monetary flows specified in the agreement have been met completely by Poland. Purchases of supplies and assemblies as well as the payments involved in the agreement itself have been made through letters of credit. The monetary transfers have been made suitable from the start. Honeywell anticipates no future problems here; as in most cases, such arrangements tend to become simplified over a period of time as transfers of payments become more routine. Most problems occur at the start, if at all.

Purchases from the Domestic Company

Purchases of total units, especially prior to the start-up of production in Poland, were generally as expected. The contract specified the prices of the complete units, the subassembled units, and the other parts available for purchase.

The Poles' desire to purchase detail parts was financially advantageous to Honeywell, since the price of these items was the same as the completed unit price. So each subassembly purchased by Poland was a source of immediate gain in its own right. Honeywell's estimation that approximately $5 million worth of products would be sold to Metronex during the course of the agreement has been a achieved.

Purchases in a controlled economy such as Poland's require long-range predictions of future needs. When these predictions also must take an unknown manufacturing schedule into consideration, such as for the class 41 transmitter, accurate assessment of potential need becomes even more difficult. It is common to overestimate the number of items that will be manufactured in the early days of production start-up. The early days of the agreement bore this out, as it was found that more unassembled and subassemblies were needed than originally envisioned. On the other hand, despite actual needs, parts in such countries as Poland could not be purchased until the money for the purchases was allocated.

Poland found that its initial estimate of need for transmitters was somewhat higher than was actually the case. However, this turned out not to be a problem, as not as many items were manufactured as soon as anticipated either. Some of the production difficulty stemmed from trying to manufacture transmitters to meet exacting specifications. For example, early transmitters off the assembly line could withstand 250 pounds of pressure, but they were supposed to be able to withstand the specified 1,000 pounds.

Personnel

Training Assessment

Any questions or concerns on the part of Honeywell regarding the ability of the Poles to manufacture the intricate transmitter vanished in the course of the operation of the agreement. In retrospect, this anxiety was well founded in that Honeywell personnel in the United States had found the maintenance of consistently high quality in the manufacture of the class 41 transmitters difficult themselves in the beginning. However, their anxiety made the establishment of correct manufacturing procedures even more important, so that the transfer would not be impeded. Thus, erasing manufacturing difficulties in the United States was an impetus to easing difficulties in Poland.

Domestic Company. The trips to Poland by Honeywell employees were just that—trips. The longest visit by any one employee lasted six weeks; no families were sent over, so any family members who accompanied U.S. employees did so at their own personal expense. In total, approximately 24 individuals from Honeywell were involved in this technology transfer, from job-related areas of design, manufacturing, quality control, and management. Generally, instruction was given to the Polish production operators, who then trained other Poles at the plant in Poland.

Honeywell employees providing on-site training in Poland apparently had no serious personnel problems. The stays were of a visiting nature and commenced for the most part when the equipment had been set up in Poland and the plant's construction was completed. At that point, Honeywell representatives went to Poland and remained until the operation was progressing correctly (which was accomplished on schedule).

Host. The underlying assumption that the Poles would have to learn all steps and phases of manufacturing and assembly of the class 41 transmitter was held by both the Honeywell personnel and the Poles. This stands out in contrast to an earlier venture with Honeywell's Scottish subsidiary, when both sides had assumed less training was needed than was actually the case. As a result, additional training and even some retraining had been necessary.

In the case of the 1974 agreement, Polish engineers and other workers were in the United States for six to eight weeks and were trained thoroughly by going over the documentation for the transmitter step-by-step. Honeywell found the Polish workers to be good, intelligent people whose aggressive and determined approach to the training provided a thorough preparation for the project.

No problems developed with security issues during the duration of the project. Various Polish workers involved with factory management came to the United States for a total of six times, while the technical personnel had a total of four visits. Some of the same individuals returned, but some changes were noted by Honeywell, especially as to the foremen. All individuals were described as highly professional and very expert in their respective fields. If party members were included in the groups, no mention was made of the fact by the Poles or the Americans. Some of the changes in individuals coming to the United States for training may have been due to visa regulations, since it was found to take two to three months to get a visa to come over for the training sessions. Visits to the United States were not viewed as jeopardizing the futures of any of the individuals involved. In fact, Honeywell noted that many of those selected to attend those training sessions were promoted or given additional responsibilities over the years.

One of the aspects of the agreement that Honeywell felt was instrumental in advancing the training methods was that those who were trained were required to train others. This helped to establish the methods in the trainer's mind as well as passing the information along to others. This requirement was specified in the agreement.

Protection

Leverage Assessment

Domestic Company. The leverage aspects that were present when Honeywell entered into the agreement remained viable, unlike other instances when the firm supplying the technology appeared to lose its bargaining position as the technology was transferred to the host country. There were apparently two reasons for the maintenance of this leverage. First, Honeywell still had a proprietary claim to the technology, which had not been duplicated by any other company, so Poland was unable to play one company off against another and had to deal with Honeywell if its products were desired. Second, since the tooling costs for manufacture of the sensor assembly were so expensive, Poland may very well continue to purchase the assemblies from Honeywell and would want amiable relations on that account.

Host. On the other hand, Poland also holds the same leverage as it had at the start of this agreement in that the country offers a highly desirable market for Honeywell products. Honeywell wants to continue to sell its products to Metronex as well as to expand its sales within Poland and other regions within the Eastern bloc, and this desire forms an element of leverage for Poland.

Guarantees appear to have been met by both sides regarding completion of their responsibilities. The only apparent issues remaining to be resolved are the terms and length of the requested extension of the agreement.

Utilized Inspections

No inspection difficulties arose, and since the products being manufactured in Poland now measure up to the specification requirements, both parties appear to feel that the transfer has been successful.

Patent/Trademark Defense

No patents were contested, and the Poles are using the Honeywell trademark as designated. Another reason for the Poles' desire to extend the agreement may be the desire to extend the use of the "under Honeywell license" trademark, and this may well be Honeywell's desire also.

Legal Provisions Invoked

There has been no need to apply any of the external protective aspects involved in the agreement thus far, such as application of the force majeure clause or resorting to arbitration.

Hindsight

Factors That Made the Project Easier than Anticipated

Several aspects made this agreement easier to complete than expected. First, Honeywell found that the skill level and the determination of the Polish workers involved with this project were much higher than they had hoped, and the technology was transferred without any major difficulties. Second, the demand for the product was such that the company was able to sell even more units than anticipated, which was beneficial to all concerned.

Factors That Made the Project More Difficult than Anticipated

Two minor timing problems developed during the course of the agreement. The first was caused by a delay in the construction and equipping of the factory in Poland and was overcome by the determination of the Poles to work overtime to keep the project on schedule. Honeywell also had leverage in this situation, since it was able to say that the company employees who were to come over and assist the implementation of the manufacturing processes would not arrive until all conditions were set for operation, which presumably strengthened the Pole's determination to get everything ready.

The second problem occurred in the United States and had to do with the development of the class 41 transmitter itself. Some aspects of its development were incomplete and had to be ironed out before the technology could be transferred to Poland. However, this was an instance when Honeywell resolved to work as rapidly as possible to iron out the difficulties and ready the project for completion, as an indication of the company's dedication to the project. As it turned out, Honeywell felt that the transfer may have actually gone smoother because of this development slowdown, since extra care was taken to overcome these problems in the transfer itself.

Progress of Future Expectations

Honeywell viewed the opportunities afforded through this agreement as twofold. First, there has been an increased level of sales to Poland through the workings of the agreement. Honeywell expected an increase, but the result far surpassed its expectations. This has included total units, subassemblies, and even replacement parts. The increase has concentrated on the area of process control equipment, with most exports coming from the Fort Washington facility of Honeywell.

Second, Honeywell is exploring the proposed extension of the 1974 agreement and feels the future holds great promise for other agreements. The desire of the Eastern European countries for hard currency provides an incentive toward manufacturing or processing agreements over the traditional export/import approach; consequently, the firms that will be most acceptable to Eastern Europe will be those whose offers take such desires into account. One surprising outcome of the 1973 agreement between the German Honeywell subsidiary and Poland in manufacturing the Vutronik line of valves is that the German subsidiary began to buy back this product from Poland—the same product they previously licensed to that country. The fact that the 1973 agreement between Honeywell, Limited, and Poland was extended is an indication that the project was so successful that the parties felt it would be mutually beneficial to extend it to take advantage of updated technology and continuing material sales.

The agreements have prompted greater interest on the part of Honeywell in Eastern Europe as a toehold for that region. An agreement is being discussed with Yugoslavia for a joint venture in computer assembly licensing. This has no direct connection with Poland, but the success and excellent relationship that Honeywell established with Poland has provided an incentive to pursue other potential contracts. All the agreements between the European subsidiaries of Honeywell as well as with Honeywell, U.S.A., have proved most successful in Poland.

Westinghouse Electric Corporation

Description Variations

The project was termed a significant success, with anticipated profits resulting from the technology transfer. Project manager, Don Walczak, reported that the Unitra plant is now in full operation under the Polish company. Walczak attributed much of the success of the project to the data processing program developed by the Semiconductor Division for control and transfer of that technology and manufacturing capability. Without such a program, he felt that the transfer of manufacturing capability could possibly have been a nightmare of delays in deliveries, documentation, or performance. Such delays could have resulted in drastic financial penalties through slipped time schedules, overruns, and inefficient use of personnel.[53]

Those in the Semiconductor Division involved with the transfer indicated that this computerized program for the technology transfer of thyristors and rectifiers to Poland was born of necessity, since a system was needed to enable the company to keep track of the various activities involved in the transfer. Walczak described the objectives of this computer program as follows:

> To outline the details of the program, define the scope, record information, effect changes, capture the technical idiosyncrasies of the process, systematize the in-

> formation for transfer, distinguish between technical needs and technical wants, define customer responsibility, control the monies allotted for purchasing, outline service requirements and minimize paper shuffling.
>
> There was also a need to key some identification system for each discrete piece of information and to be able to rapidly communicate any changes in information that would affect performance of the project to the customer or purchasing within the time scope allotted for the project. It was with these objectives that the program was formulated.[54]

To accomplish these objectives, Walczak set up a three-part program. At the heart of the program was a numbering system for each item—for instance, every item was given an eight-digit number, two digits reflecting its location to a work station, two for its operation, two for its function, and two as its own special identification. Three types of printouts were used: bills of material, financial information, and services information.

Through Walczak's efforts and strict adherence to his program, the project in Poland functioned on schedule and was accomplished in perhaps the most systematic way possible.

It was found that the original contract did not cover all the issues that arose, but addendums were added to apply to such issues as assembly line yield, processing yield, and voltage design. To solve the voltage problem, Westinghouse adapted its product to the Polish environment by designing a unit that would operate with 2,800 volts. Since the Semiconductor Division had no other market for this product, this was just one indication of Westinghouse's concern toward the project. Much of the ultimate success of the venture has been attributed to the efforts of both parties.

Systematic control over each step helped to prevent confusion and waste. For instance, when equipment was received in Poland after having been tested and examined in the United States, the items were signed off as they were taken from the crate; any missing items were immediately noted or requested at that time. The equipment was placed in locked bins until needed, but this precaution was apparently unnecessary, since no theft was ever reported. All major items were immediately placed in the proper location, so there was a minimum of confusion in the unloading or unpacking of crates.

Walczak noted that the purchase of all new equipment for this project was an advantage for Westinghouse. Not only did the corporation first have the opportunity to systematize the shipments and pretest the equipment, but, in many cases, start-up service for the machines was one of the conditions of sale. Westinghouse functioned more as an intermediary or agent in the purchase of machinery and was able to stipulate conditions when buying the equipment and machines.

Westinghouse found the creation of charts and signs designed for high visibility in the workplace very helpful during the transfer. A system of buzzers, lights, clocks, and recorders accompanied the charts. The objective was to have each step so visible that details that might have been overlooked on log charts or in paperwork could not be overlooked during implementation. Thus, a manager had only to glance up to see the progress that was being made. This system was used during the installation phase of the operation and in teaching the manufacturing steps. Product flow was designed to move in the form of a three-leaf clover, or from storeroom to storeroom. When the three loops had been completed, the product was finished. This provided an accurate means of controlling product quality, since steps that were being inadequately completed could be pinpointed and corrected.

Although the contract did not specify the transfer of any general plant management information unless it pertained directly to the manufacturing process, Westinghouse found that such training was needed. Initially, general management training was limited to product or inventory control, but it evolved into more than that. Westinghouse found that the Poles needed assistance in general procedures, such as how to move the equipment in most efficiently. If rollers or dollies were not available, they were constructed. Cleanliness is crucial in the production of semiconductors because quality control is so critical, and soon Westinghouse was teaching standards for cleaning, vacuuming, and painting in order to reduce the dust in the facility. A group of university students were even hired to scrape the windows so the glass was clean.

No changes occurred in the patent or trademark associations provided to Unitra. However, one final, rather unique task done by Unitra was to create a film of the project including views of the site, the plant, and the various operations. This task may not be unusual for some projects; however, because the security within Eastern Europe is more restrictive than in some areas of the world, it was unique for this type of operation. It was also highly beneficial to both parties. It provided Unitra with a tangible record of its semiconductor operation and sparklingly clean factory, and it provided Westinghouse with a film version of its successful conclusion to the project.

As expected, the patent portion of the agreement was extended for three years, permitting Unitra use of its Westinghouse association on its license indication. As before, no direct trademark usage was permitted, and once the extended patent agreement expires, Unitra will not be able to use the Westinghouse connection in its sales promotion or on its products.

The teamwork approach was the work of both parties. As noted above, Westinghouse did help Unitra by providing some training functions that had not been specifically listed. Westinghouse also pitched in at various times to help Unitra achieve or complete some prime responsibilities if it became apparent that some assistance could be helpful.

Current Corporate Perception of Benefits from Agreements

The actual outcome of the Polish agreement was apparently in line with Westinghouse's expectations and reasons for desiring the contract. The project's completion without the incurrence of any additional expense justified the major reason: a transfer of technology for the fees obtained from that sale. Essentially, the fees paid for the technology functioned as royalties, which may be viewed as comparable to the profits that would have been obtained had the product itself been sold there. While the technology surrounding power semiconductors may be updated rapidly, the technology sold to the Poles will not be obsolete for some time, especially as applications are plentiful for silicon-controlled rectifiers in Poland.

Westinghouse's second reason, a toehold in Poland and throughout Eastern Europe, had not materialized by late 1979 to the extent desired by the corporation. One possible venture, an agreement for the production of color television tubes, was awarded to the RCA Corporation and Corning Glass Works in late 1976. However, even though no agreements had been concluded, a far greater awareness of Westinghouse and the corporation's products existed as a result of this project.

Furthermore, this awareness developed long before the project was completed. Westinghouse had anticipated that any further contacts throughout Poland or Eastern Europe would be as a result of the sale of the semiconductors produced at the Lamina facility. However, it became apparent that the corporate reputation spread as a result of the project itself, long before the products reached the market. In fact, Westinghouse was invited to give a technical presentation in Romania just because word of the corporation's involvement had spread to that country.

It is often pointed out that personal friendships are very important in Eastern Europe. This agreement was a prime example. Project Manager Donald Walczak virtually commuted between Pennsylvania and Poland from 1974 until production was under way in 1976. Romanian officials had not heard much of Westinghouse but they had heard of Don Walczak. On the basis of his reputation, they requested a presentation to find out what Westinghouse was all about.

Westinghouse officials felt that a joint venture would probably be needed in Poland to actually provide a proper foundation for future contacts. The corporation's facility in LeMans, France, was apparently concerned about the Unitra project, viewing the operation in Poland as a potential competitor. Since LeMans possessed both the Westinghouse trademark and a highly active sales force, the competition even at this point in time is hardly severe. Westinghouse reiterated its hope that the success of the Polish venture would provide assistance in obtaining future contracts throughout the Eastern bloc.

The third body of advantages sought by Westinghouse through the contract remains potentially viable although, so far, none have transpired. Included in this category is the possibility of another source of semiconductors for Westinghouse on a buy-back basis and access to raw materials in Poland. No additional semiconductors were purchased from Westinghouse prior to the start-up of production at Lamina (over and above those used as training aids), but such sales were not anticipated.

From Unitra's perspective, sales of the semiconductors remain a potential source of hard currency for Poland, but one that has not been greatly tapped as yet.

An unanticipated benefit from this agreement was the broadened perspective gained by the Semiconductor Division in the United States as a result of breaking down the entire transfer process into a detailed computer program. The transfer became a highly controlled experiment that provided new insights into division design, processing, and other methods of production. The documentation of its know-how provided information that brought more order and methodical control to the division and heightened the future use of its resources.

Negotiations

Contract Format Revisions

The contracts comprising the agreement between Westinghouse and Unitra were apparently quite suitable. Fifteen additional addendums were constructed as needs arose. Most related to specific issues requiring relatively minor revisions, such as manufacturing or equipment and tooling information. A protocol was also written regarding the final acceptance tests (November 1976), detailing what and what not had been accomplished, even listing spare parts that were still missing. Westinghouse noted that all funds set aside to purchase equipment had been spent.

Westinghouse judged its earlier negotiating sessions with the Poles as demanding but fair. "The Polish people are hard negotiators, and they held the line hard at getting what they paid for, but there was no attempt to go beyond that," was Maurice C. Sardi's opinion, former Semiconductor Division general manager.[55]

Westinghouse was never able to learn whether it had been in direct competition with other U.S. companies for the Unitra contract. Westinghouse had no information from its own sources, and Unitra was not willing to divulge this information.

At the conclusion of all negotiations, the final agreement had been handwritten in Poland then typed there in English. Both parties had ample opportunity to carefully analyze all portions, and both sides felt fully cognizant of all clauses of the contract. Apart from the letter that kept possible buy-back arrangements open for Westinghouse and the addendums, no major changes were added to the contract. The patent provision was accepted for the three-year extension, so the patents will be valid through 1981. This actually applied to only one patent, but it enabled Unitra to use the phrase "Made under license from Westinghouse" until the end of that time period. No additional fee was involved for the patent extension.

The updating feature of the five-year agreement was not extended. This involved the continuing transfer of all relevant improvements for the particular thyristors and rectifiers purchased by Poland that might be developed by the Westinghouse Semiconductor Division during that time period. This service concluded in 1978.

Pending Negotiations

As of 1978, no additional agreements had been concluded between Westinghouse's Semiconductor Division and Poland—or anywhere else in Eastern Europe, for that matter. However, Westinghouse does not rule out future agreements, feeling that the successful conclusion of this agreement will enhance all further contracts. As a result of this successful project, the company is looking into the establishment of a similar plant in the People's Republic of China. Westinghouse is confident that its systematic approach to technology transfer makes such interchanges even more desirable from both buyer and seller standpoints.

Time Frame

On Schedule

The agreement officially concluded on April 1, 1979, but this was almost an anticlimax in terms of the project's development.

The crucial deadline in the contract was the November 1976 deadline for the acceptance tests, and this deadline was met on schedule. All necessary tests were passed. In Unitra's view, this indicated that production was on-stream for the power semiconductors that were needed throughout the Polish economy. In Westinghouse's view, the lack of delays meant that the corporation's profit from the fixed-price contract was secure and free of expensive delays. As anticipated, the patent portion of the agreement was extended for three years, or until 1982.

Behind Schedule

Even though the final deadline was met, several incidents occurred in the course of the 32-month setup period that might easily have disturbed the efficiency of the project. During the preparation stage, all equipment was assembled at the Youngwood division's plant for shipment to Poland. This was a major factor in maintaining the schedule. However, in Poland the situation was less efficient. Any materials that had to be procured there had to be ordered through a procurement-type company, accompanied by three quotes. This lengthy and time-consuming procedure resulted in many delays. The building shell was complete, but the supplementary materials needed for the renovation were difficult to obtain. Setting and maintaining certain standards as to cleanliness within the plant required additional work on the construction and some installation. Once these problems were surmounted the schedule appeared to progress as anticipated.

Financial Flows

Payments

All payments were received from Unitra as scheduled, with the final payment arriving by December 1976. The extension of the patent agreement for three more years did not require any additional payments on the part of Unitra. By timing the payments to correspond to the different work segments, Westinghouse found a basis for accomplishment had been provided, as the payments then visually marked the progress of the agreement.

Most contracts within Westinghouse's Semiconductor Division are not as large as this project in Poland. Hence, most do not require such complicated financing arrangements involving the Eximbank as well as other U.S. commercial banks. Time and effort were required on the part of Westinghouse to explain the financing arrangements to both the treasury representatives within Westinghouse and to the outside banking officials involved on the project.

The acceptance tests were completed as scheduled. This is particularly important to a corporation's overall profit projections in a fixed-fee type of contract such as this one. Additional testing would have been at the expense of Westinghouse, unless the problem had been with Unitra alone.

Purchases from the Domestic Company

The equipment purchases went as arranged. No additional assemblies or subassemblies were requested beyond those used for training.

Westinghouse had a surprise regarding service or replacement parts. The contract stipulated that spare or replacement parts were to be either kept in reserve or available. However, because the demand for spare parts was so high, Poland purchased more spare parts than Westinghouse ever anticipated. When possible, the spare parts were purchased from the Westinghouse LeMans operation in France.

An indirect consequence of the agreement surfaced, involving the equipment suppliers used by Westinghouse. Since Westinghouse acted as the prime contractor for much of the equipment installed in the Lamina factory, the vendors used by Westinghouse were introduced to Eastern Europe. The reputations of these suppliers increased as a result and they were able to sell to other Eastern bloc countries. Many were invited to exhibit at some of the trade fairs, for example, in Munich.

Purchases by the Domestic Company

As of this writing, Westinghouse had not exercised its buy-back option from Unitra because it had not yet encountered a need to purchase any of the items manufactured there.

A number of small but important improvements were required during the construction and equipping phase at the Lamina facility. The water system had to be improved, if only to maintain a constant pressure; otherwise, even the toilet tanks would not operate. Additional lights had to be installed. Instructions at the work stations were printed in both Polish and English to speed comprehension and avoid mistakes. Unexpected tasks were accomplished through hiring part-time workers. For example, students were hired to scrape and clean the windows.

The expenses incurred by these improvements were covered by Unitra, but it is important to realize some unanticipated expenses will usually arise on any project and provisions should be made in advance regarding such costs.

Personnel

Training Assessment

Domestic Company. The Youngwood representatives on the Unitra project included Paul E. Lego, general manager; Thomas P. Nowalk, advisory engineer; Donald Walczak, project manager; engineers Larry D. Hall, Richard Harm, John P. Lindenberger, Ronald P. Nandor, and Edward D. Pomper; and draftsman Richard M. Beabes. Engineering support services were provided by the Westinghouse Manufacturing Development Laboratory in Pittsburgh. There were approximately 16 people from Westinghouse involved in the project, with 11 individuals on site. This number generally coincided with the anticipated size. However, more than 80 round trips were made to Poland (72 during Phase III alone), which definitely exceeded expectations. It was felt, however, that the trips were definitely needed, and Westinghouse absorbed the cost of this additional travel. The length of an individual's stay was not indicated in the original contract, but Walczak made it a policy that no Westinghouse employee remained longer than six weeks in Poland without a two-week break back in the United States. No Westinghouse employees remained in Lamina for the duration of the project. No families were moved to Poland. True to predictions, the acceptance tests were completed by November 1976, and all members of the Westinghouse team were home by Thanksgiving, except Walczak, who left November 31, 1976.

The skills possessed and used by the Westinghouse employees fit the tasks at hand, because the employees were not sent to Poland unless the work there was ready for them. Work was adapted to the individual, and this was done for both the U.S. and Polish personnel. In addition to the technical and engineering expertise of the Westinghouse employees, it was also necessary to pass along their managerial skills. Westinghouse personnel were involved in the building renovation, too, something not anticipated beforehand.

Basically Westinghouse kept to the budget laid out for the project. The two major expense areas that were not specifically anticipated included car rentals (to reduce the time spent waiting for buses) and the additional trips.

The Poles were responsible for providing "available suitable" housing for the Westinghouse team, but the apartments initially selected were not deemed fully satisfactory. Instead, Westinghouse employees moved into a small hotel near the Wisla River. These accommodations, paid for by the Poles, turned out to be ideal. The hotel provided a charming gathering place as well as a homelike atmosphere, and the personal service was excellent.

There were very few actual problems. Some harassment by the local police was noted during the first two to three months, but this quickly died down with no repetitions.

Some Polish working habits surprised the U.S. employees. For example, the Poles did not take lunch breaks. The Polish workers would have an early breakfast before starting work and take a second breakfast about 9:00 in the morning. This was followed by frequent coffee breaks until the workers went home at 3:00 in the afternoon. Following the philosophy of being open-minded about local customs and adapting to that style whenever possible, the Americans soon brought out their Mr. Coffee machines and stocked their offices with peanut butter and other foods. This way, in the spirit of "when in Poland, do as the Poles," they were able to take short breaks instead of a lunch hour.

Westinghouse approached the conclusion of the project with the same care and planning it had followed throughout the entire period. Project Manager Walczak stressed the end points. By sending the Westinghouse engineers home, the Poles were encouraged to operate the factory on their own. Walczak noted that it was hard to let go after two to three years of close association, but he realized that the Polish technical people had to be put on their own. The Americans were physically removed from the site gradually, over a period of time. Finally, Walczak was the only American remaining. His final duty—after completion of the protocol for the successful conclusion of the acceptance tests—involved listing any "exceptions," such as things not done or spare parts not delivered.

<u>Host</u>. Westinghouse felt that the success of the project was due, in part, to the high caliber of Polish people involved with the project. General Manager Lego observed, "(The Polish Government) assigned talented people to us, and the work proceeded in an atmosphere of mutual confidence and trust. There was never any feeling that anyone was trying to put something over on anyone else."[56]

Westinghouse found that the Poles had excellent training in chemistry and physics, with specific skill concentrations almost to the point of having tunnel vision. It was up to Westinghouse to provide a broader overview, relating these individual disciplines to the production of semiconductors. The company found this much more preferable than situations in which "unlearning" must be done before "learning" is possible.

Some 50 hourly workers and 25 professionals were employed in operating the new semiconductor facility, but over 200 Poles were involved in the total project, including the construction and installation phases.

The training sessions in the United States varied from three to ten weeks depending on the type of training. Six Poles were sent in each group. All 50 man-months allotted for training were completely used, but no additional training time was required. In general, different individuals were sent over with each group, but the same individual would return for additional training on rare occasions. The Polish technical manager was the acknowledged leader for the group.

Westinghouse set up an individual training program for each person and kept notes on each session. The company also arranged all training so that not only were the most appropriate instructors there but also the right sets of tools. The training sessions generated a great deal of rapport, especially with the project engineers.

Westinghouse used the management team concept in the training to implement the concept of the processing center, with areas for slicing, lapping, and diffusion. Thus, the teams were exposed to all aspects of the manufacturing process.

No unexpected expenses were generated or created throughout the training sessions in the United States. Poland paid all financial commitments for training, as designated.

Any problems stemming from the Poles' training were relatively minor and certainly overshadowed by the opportunity that Westinghouse had to work with such able, dedicated workers. Therefore, the following observations are more accurately termed information, not problems. All Polish trainees sent to the United States for training apparently did not have the same level of involvement with Unitra. Some individuals were key technical personnel, others came in a supporting capacity. Only minor language problems existed. All trainees took classes in English pertaining to the technical areas involved in semiconductor manufacture. In like manner, all U.S. engineers involved in the project studied Polish for a year before going to that country, with lessons two and three days a week. It was noted that there will usually be language problems when dealing with technical areas.

Other Personnel Involvement

Some management rules in Poland differ from their U.S. equivalents. For example, any employee who became ill automatically received three days off. As a result, absenteeism was relatively high, and the Americans never knew whether an employee was going to be absent or not.

The contract stated that safety rules had to be followed. But it was difficult to enforce those rules universally when the regulations appeared meaningless. Westinghouse attempted to provide the reasons behind regulations. One instance involved some spilled acid that could have resulted in burns and perhaps blindness. Fortunately, the employees were wearing safety glasses and safety shoes. The company found it best to concentrate on realistic regulations that were not overly protective.

The morale of the Polish workers was quite high. They exhibited a great pride in their country and in their technological improvements. They appeared to like the United States, but seemed somewhat apprehensive of the competitive nature of the capitalistic system, preferring their secure system instead.

Interestingly enough, Westinghouse found itself providing marketing expertise as well as management training during the course of its contract with Unitra, although this was also beyond the specific scope of the project. This advice benefited all parties. If Unitra sold more semiconductors, both Westinghouse and Unitra benefited from the actual funds and the increased reputation.

Protection

Leverage Assessment

Domestic Company. Westinghouse maintained its initial advantages throughout the course of the agreement. This continuing leverage apparently had several contributing sources. First, the agreement was contractually fulfilled when Westinghouse met all corporate obligations. In connection with this successful completion, Westinghouse established personal contacts and friendships, providing an excellent basis to further relationships in the Eastern bloc.

Second, despite the five-year updating period, new advances would prompt desires for extended technological assistance agreements. This, combined with Westinghouse's growing reputation throughout Eastern Europe, made both the company and its products more desirable. The firm was not seen as the sole source of most technology, but it was viewed as one of a small number of firms possessing excellent technological resources.

Westinghouse illustrated the company's leverage on the Unitra project by not shipping equipment before the plant was ready. Trips made to determine the status of the project were done before approvals were granted. However, no crucial delays occurred, and no delays longer than two months were encountered. These created no major problems, partly because the contract schedule had allowed for some delay time.

Host. Poland's leverage was also maintained throughout the contract, although, as with Westinghouse, the greatest proportion of specific responsibilities occurred within the first 32 months of the agreement. Although Westinghouse views future, similar alliances with Poland as desirable, Poland still holds the power of awarding those future contracts.

Poland's leverage with Westinghouse also stemmed from the contract itself, just as Westinghouse had similar controls with Unitra. One example would be Poland's request for a final protocol, which formally listed all complete and incomplete aspects of the transfer at the time of the official turnover to Unitra.

Westinghouse met all guarantees as laid out in the agreement by successfully transferring the manufacturing capability to produce power semiconductors. Likewise, Poland met all of its guarantees, which were vital to the success of the operation in both the construction-equipping area and in the training area. In both instances, it was felt that the guarantees were not really necessary for the project to work, but they were desirable legally. As it turned out, these guarantees were not utilized in a legal sense.

Utilized Inspections

Westinghouse was permitted to examine the facility, although the U.S. personnel were always escorted when moving from one building to another at the Lamina complex. Technically, the company's right of access ceased upon completion of the facility, but the rapport was so high by that time that Westinghouse was under no obligation to leave. The firm's actions in departure were designed to get the facility totally under Polish operation in a training sense, not for any other reason.

Patent/Trademark Defense

No patent or trademark disputes or infringements arose.

Legal Provisions Involved

No instances activating the force majeure clauses arose and no disputes were taken to arbitration.

Hindsight

Factors That Made the Project Easier than Anticipated

As a matter of corporate policy, Westinghouse analyzes each transfer project to study methods for improving its transfer system. The analysis of the Unitra transfer concluded that there were three reasons for the success of the project. One, the product, as it was "frozen" in its 1973 state of development, was more than adequate and therefore remained substantially unrevised, with a few exceptions. Likewise, two, the process, which was also detailed in its 1973 state of the art, was excellent. Warnings were given, however, that all processes had to be followed to the letter. The third reason was that personnel, from both Westinghouse and Unitra, achieved high levels of rapport. The Polish manager of the project was singled out as an extremely excellent person. If assistance was needed, he knew where to get it. Westinghouse also found that its project manager, Don Walczak, and the rest of the project team did a superb job. The fourth element contributing to the project's success was that the planning—particularly the systematized, computerized approach, supported as it was by organized filing, record keeping, and even the telex system—provided a comprehensive scheme for implementing the transfer process.

Several other points, while not responsible for the total success of the endeavor, certainly played supporting roles in making the project easier. One of these was the creation of the film, which provided a stimulus for concluding the endeavor and a reminder that the Poles were able to manage the plant. Another was the ability to pinpoint areas of difficulty—actually part of the planning scheme mentioned above—which not only helped Westinghouse improve the transfer to Poland but also enabled the company to examine its own process to make it more efficient and to provide a better assessment of its own operation. Westinghouse also pointed out the advantages of using new equipment of high quality. The operation could then be standardized and any initial problems corrected within the warranty period, another boom to keeping profit margins close to the anticipated level. The inclusion of installation and start-up as part of the purchase price was also beneficial.

Westinghouse used a system of accumulating all equipment, including all necessary tools and instructions, at the plant in Pittsburgh and then sending these items via container to Poland when the time schedule indicated that installation would be relatively soon. This provided a control over the equipment that minimized loss, breakage, and operation mistakes.

The emphasis on visibility was important in controlling the entire training process. Charts on the walls, with sound apparatus warning of problems, were read more easily and rapidly than reams of written reports—particularly in a situation where dual languages were involved.

Finally, an element that helped the entire project was that progress projections had allowed sufficient time for all operations to be completed within the time frame. Delays did occur, but provisions had been made for possible delays so that the operation was not jeopardized.

Factors That Made the Project
More Difficult than Anticipated

Possibly the most crucial factor was the outages that occurred in Poland regarding some raw materials, particularly metals. The time and effort required to locate suitable substitutes created lengthy delays. Testing raw materials for specifications was also time consuming, as was making provisions for the management training of Polish supervisors.

Restrictions in Poland also contributed to delays, although these could have been far worse. One example of these delays was the four-to five-month period required to obtain permanent (until the completion of the project) work visas.

Progress of Future Expectations

None of the possible contracts between Westinghouse and other countries in the Eastern bloc had materialized by 1979, although contacts had been made. Westinghouse did exhibit its wares at the trade fair in Romania, and other countries were still expressing interest. An arrangement with Romania was being pursued but had been postponed for further consideration. A possible arrangement between Westinghouse and the People's Republic of China is viable, as contracts have been made with visits and technical discussions in the offing.

The rapport between Westinghouse and Poland is expected to lead to other important contacts in that region, as the two parties are looking for areas where mutual needs may be satisfied. Presently, Unitra requires a stronger financial basis before being able to pursue additional contracts with Westinghouse.

Westinghouse's sale of technology continues to be a viable approach. The corporation's technology sales have increased proportionately since 1973, substantiating this opinion. Such rewarding ventures as the Unitra-Westinghouse agreement have served to strengthen this position.

ROMANIA

General Tire International Company

Description Variations

After six years of operation, the agreement between General Tire International (GT) and Romania had accomplished much. However, a number of unexpected situations arose during that time, far more than General Tire, with its wide background in overseas ventures, ever anticipated. This is not to say that all the occurrences were negative in scope; but it must be pointed out that several unusual events transpired in those first six years.

Perhaps the best overall summary of General Tire's problems in Romania was given by Gustav Prekopa, GT's coordinator for the Romanian Factory Project.

> General Tire International has performed this routine of technology transfer in many various countries time and again successfully, but never have we earned our fruits of labor as thoroughly as in this project. What is then this basic difference that we encountered for the first time in Romania that made our task so much more complex? I don't think it can be really quantified other than with a vague generalization. One difference which is apparent is that individual enterprise, so prevalent in the West, does not exist in Eastern Europe in the same manner. Instead, individuals conform to the State-administered system. Individuals do not make decisions—the group decides. We have found that the individual tire technicians in Romania are extremely knowledgeable, highly qualified experts, who have adopted and assimilated the training rapidly and thoroughly. (But,) everything has to be committed to writing which, in turn, can be then analyzed by a collective group for it to formulate an opinion, an acceptance, or a refusal. This arrangement in practice makes technology transfer considerably more difficult and frustrating.[57]

One of the greatest frustrations GT encountered appeared to be a lack of inclination or inability—for one reason or another—to assume responsibility. It was also a problem that GT had not encountered prior to this project, as they stated that "outside the Eastern Bloc, in East Africa or in the Middle East (the corporation had found) a willingness to share the blame."

Once this problem was recognized, it was possible (at least at times) to allow for such behavior and take appropriate action. It was noticed that as long as a decision did not have to be made by a Romanian, all was fine. Actually, GT found that the problem was not an inability to assume responsibility, but rather that the workers involved with the project were just not permitted to make mistakes. Having to make a decision included the possibility of making a mistake, or possibly having to admit to having made a mistake. Thus, one method used during construction to circumvent decision making was to have a daily log sheet listing what was to be done. The U.S. supervisory staff went so far as to make chalk marks on the floor where the machines were to be installed, eliminating even this level of decision making for the Romanian workers.

There was a positive side, too, to this responsibility issue that became evident in the Romanians' approach to their machinery or even total plant purchases. It was noted that the Romanians wanted the best in machinery and the latest in technological developments—the "gold-plated factory" concept. The initial price for such items might be high, but the technology associated with them would not become outdated as rapidly, and the machinery would tend to run better and longer, even if the maintenance on that equipment was not as thorough as desired. A GT supervisor in Romania pointed out that the plant's machinery was operating better than ever anticipated—especially in light of the improper oiling and maintenance schedules. One GT official noted that such gold-plated factories would produce goods at a higher, more constant level, thus providing a more constant measure within keeping of the dictates of the five-year plan.

Current Corporate Perceptions of Benefits from Agreement

The initial reason for GT's interest in the Romanian project was the financial gain to be realized through the sale of technology to that particular country. It is difficult to assess whether GT's desire to expand throughout Eastern Europe was more the cause or the effect of its interest in Romania. It could easily have been more convenient and coincidental than premeditated, in this case. However, because the company is so internationally oriented, the prospect of a toehold in the market was certainly one factor, if not the major one.

GT's expectations that Romania would become a source of scarce raw materials has not materialized, but the corporation's continuing involvement in the country is the basis for its hope of gaining some of those raw materials in the future.

On the other hand, with the demand for radial tires growing throughout Europe and the Middle East, GT saw the production in Romania as a source of supply of the completed product, which the corporation might then be able to use through repurchases to help its own sales internationally. This possibility is often present in agreements in which General Tire provides its trademark or logo, but the increasing demand for tires combined with the high quality of the tires being produced by Romania (or, expected to be produced by Romania once the acceptance tests have been completed) has magnified this expectation in this particular instance.

Negotiations

Contract Format Revisions

The agreement between General Tire International and Romania was awarded and became effective on March 1, 1973, but General Tire found that the conclusion of the contract discussions was really only the start of a bigger round of negotiations. Since that date, 25 protocols have been signed between GT and Romania—amendments that either altered or clarified aspects of the contract. The intent behind structuring a protocol is to introduce certainty into uncertain situations; it is a midcourse negotiating device. In this instance, one party apparently started reacting to aspects of the contract that were particularly beneficial for the other party but not for both. GT found that, based on its experience, the negotiations had only just begun when the contract was signed. The entire project has apparently become one long, drawn-out negotiation.

GT's recommendations regarding postcontract discussions are the same as those developed during the initial negotiations: everything should be documented. One official stressed the concept that although both parties might appear to be in agreement, nothing is certain until that agreement is down in writing. The corporation found that its best negotiating position was to be able at any time (if possible) to terminate performance with a minimum of financial loss—or, as verbalized by one official, "don't get yourself in so deep that you can't walk away."

GT found the entire negotiating process far more difficult than anticipated. It was expected that the negotiations for any contract would require at least a year, but GT greatly underestimated the red tape involved in dealing with a country in Eastern Europe. However, it was stressed that the "unbelievable" red tape was not the fault of the Romanians as much as it was the entire problem of dealing with a socialist form of government in which everything has to be exactly laid out ahead of time.

At times, the GT negotiators felt that the Romanian position was unreasonable, but GT was being forced to try to find a common ground somewhere between the Romanian position and GT's realistic one. These situations occurred, it was stressed, regardless of the topic—whether the session involved the collection of funds, the performance of work, or a discussion of difficulties.

GT sometimes felt that even the structure of the Romanian language was designed to put the supplier, General Tire, on the defensive. (This effect may have been intensified by the translations.) On top of this, each work was taken at its literal meaning and the contract was viewed as "the law."

Additional care was required to thoroughly peruse each section of the contract and its revisions, as the contract language was often rephrased. During the negotiations, portions would be lost or varied —for example, the statement "delivery penalty will be three percent of the undelivered portion of the value of the subsection," might become "delivery penalty will be three percent of the value of the subsection," resulting in a sizable increase in the penalty.

Another slipup resulting from the rephrasing process involved "the work required to be performed," which was reduced during the negotiations, along with the price—but the number of on-site man-days required of the supplier was not proportionately reduced, thus, remaining the responsibility of the supplier who must then incur additional effort and expense.

General Tire found that one way to restore balance during negotiations was to contribute additional items to the contract settlement—for example, making sizable spare parts or supplies contributions, whereupon the Romanians would permit contract language interpretations to be modified.

GT observed that part of the difficulty in dealing with Romania lay with the underlying cultural tendencies. These more closely resembled the Latins or Italians than other northern Eastern Europeans. This was reflected politically, too, in the Romanian's almost feudal dedication to the president of that country. In their form of government, which tended to resemble Stalinistic communism, the word of the government was absolutely final. General Tire had expected more flexibility, despite a prior knowledge of their system of government.

The technology transferred caused problems, too, for several reasons. The Romanians wanted technology for items outside GT's realm. The items were connected with GT's authorization in providing such technology from outside suppliers.

The Romanians also criticized the technology that GT was able to transfer. For example, one person involved with the project pointed out, "They still want a better blend rate than our own technology calls for." The technology transferred was GT's latest information on radial tires. GT did not want to promise more than it was able to deliver, and it wanted to accurately portray the finished product so that the Romanians would not come back and say, "You have given us a product inferior to the one you promised." On the other hand, GT certainly did not want to downgrade its technology or provide descriptions that fell short of its capabilities.

Some of GT's problems can be attributed to the contract's wording—for example, if something was not specifically excluded, it was, by default, included. Consequently, GT became involved in training the Romanians to use the new equipment, something that had not been anticipated. Because it had not been specifically negated by the contract, it became one of GT's responsibilities.

"Normally in a contract," one official stated, "you don't have a thousand little details. Eastern Europe, however, wants a huge Pert Chart, measurably pinned down." But, after they have such a chart, another GT official noted, "they don't need you any longer—they'll do it themselves."

Communication, over and above language differences or the translation between the two languages, was difficult. The underlying problem was that the Americans felt that the Romanian style of communication was always threatening. The Romanians took a firm position for every encounter and expressed that position with force, regardless of the necessity for such an approach. For example, GT discovered that a statement such as "you are in violation of Article 5.x of the contract" might actually mean "we haven't received your invoice to permit the time-based payment due under the contract." Or, another example, the statement "you have not fulfilled your obligations to provide complete documentation" might mean "we can't find the anchor bolt description required for the item of equipment."

In addition to the defensive nature of the communication style, the contract language was rigidly enforced, regardless of the logic behind the statement or the situation. Thus, problems with the respective languages were apparently not from translation difficulties, but more from interpretations of the content.

Looking back at the situation, GT noted that it might have been better to have had an equipment supplier as a buffer company. GT could then have furnished a list of equipment and a description of the necessary types of technicians required to install it. The equipment supplier might even function as the prime contractor, even if the specific technology transferred originated with General Tire.

In retrospect, GT noted that the contract format might have inadvertently fostered frustration by having too many areas that needed interpretation. Furthermore, GT realized that while the contract appeared to specify the tasks that its company was to perform, Romania's responsibilities started and stopped with paying for the project. Romania was not specifically responsible for accomplishing certain duties, only for paying for them. If GT were to draw up a new contract, it would specify the responsibilities of both partners, in addition to payment schedules.

Pending Negotiations

Romania indicated a desire to acquire GT's technology for off-the-road vehicles. Ideally, this technology would complement Romania's current technological package from General Tire. However, the corporation had adopted a policy of waiting until current projects had passed their acceptance tests before considering further involvement.

Time Frame

On Schedule

The project appeared to be behind schedule for several reasons.

Behind Schedule

General Tire found that its expectations regarding the timing of the Romanian project had been too ambitious for several reasons—some within the control of the company and some totally outside its jurisdiction. After five years, the observation was "unless you have the patience to take three weeks to accomplish three days work, don't go."

In retrospect, GT felt that its time frame expectations had been overly optimistic, but part of the reason for setting the 1975 construction completion deadline (with the accompanying start-up in production) was the Romanian desire to accomplish this by the end of the five-year plan that concluded in 1975. The Romanians wanted the production under way by the end of that calendar year. Even at the beginning, GT felt that the construction deadline for such a large plant should more accurately be outlined within a time frame of three years. In this case, completion should have been scheduled for March of 1976.

Design and equipment delivery took the anticipated 18 to 20 months and continued while the plant was under construction by the Romanians.[58] General Tire did face equipment delivery problems due to the petroleum shortage caused by the OPEC oil embargo in the United States in the early 1970s. This indirectly (if not directly) slowed down construction in a number of areas for the years 1974-75. GT had advance-ordered the major machines needed for equipping the Romanian tire factory, since such items had to be custom-built to plant specifications (from GT prototypes). However, due to the recession and the resulting slowdown, it took almost five years to totally equip the Romanian tire factory, and virtually another year before manufacturing lines were installed and equipment finely tuned.

General Tire felt that the plant's construction went very well—considering that the Romanians were adapting tire production to GT's instructions, just as GT was adapting to Romanian life-styles.

Training, both on site in Romania and at several GT locations in the United States, was in progress throughout the construction and installation phases, so this aspect was not completely upset by the construction and equipment delays.

The first tire came off the line in June 1975, 30 months after the signing of the contract—in record time. However, this was only part of the pilot run, since the equipment was still in the process of being installed. By the end of 1976, production was up to 60 percent of the projected total capacity; total production was achieved by March 1977. Unfortunately, during that month, a major earthquake caused extensive damage to the plant and it was another 12 months (or March 1978) before production resumed in quantity.

The earthquake was totally disruptive and exerted a devastating impact. However, it was not the only event to delay the acceptance tests, which had to be performed on the tires before final approval could be given. Another problem involved the test vehicle that was being used to compare the tires produced in Romania with tires manufactured by GT in the United States. In April 1978, during one of the extensive road tests, the vehicle went off the road, incurring extensive damage and suspending further tests. Another road test was begun in June 1978, but since these tests must run for months it would be at least June 1979 before this test could be evaluated. This road test had to be satisfactorily completed before the protocol could be signed, which had signaled that the tire quality was acceptable and in accordance with the standards laid out in the contract. (Protocols were the official documents that verified the progress of the contract; they did not change the intent of the contract. Several types of protocols were used on this project. Descriptions accompanying major protocols were termed sight protocols and were used to clarify or annotate certain portions. For example, one sight protocol laid out the route for the final road test.)

Another timing problem occurred in the final stages of plant construction. GT had had difficulty getting some of the earlier protocols accepted, which had to do with the completion of certain aspects of the plant. Those on the scene, both the Americans and the Romanians, were having no success. The deadlock was finally broken only when higher Romanian officials were brought in who were in a position to make decisions regarding the completed stages. These officials were appointed to keep the contract going in Romania, since they possessed both the authority and the ability to work with the local managers at the site.

General Tire would restructure a number of aspects of this project if it were possible to restructure such things, but unexpected calamities such as the 1977 earthquake would still go beyond normal revision projections. On the other hand, the fact that such a disaster did occur only strengthens the statement that unexpected occurrences need to be built into a project's time frame and cost structure.

The extended time required to complete the acceptance tests was a financial detriment to GT in several respects. First, the additional revenues from the installation of the technical service contract would not commence until the acceptance tests had been completed. Furthermore, since no escalation percentage or fee had been attached to the contract, inflation steadily reduced the amount of real dollars that GT would gain from this service contract. Second, the delays resulted in greater expenses for GT in the form of additional personnel and attention paid to the project. Third, since the contract did not forbid production of tires in the plant prior to the acceptance tests, but only forbade use of the GT official logo, the machinery was being operated and tires were being produced and worn down before the tests had been completed. This, in turn, was forcing GT to prove contract acceptance on machines that had been worn down, potentially jeopardizing the tests and extending the time period for completion of the tests.

GT's comments regarding the culmination of the acceptance tests for tires from the plant were based on its experiences over the past six years. One observation was that "acceptance tests on individual machines will succeed or fail depending on whether operation of the equipment involved makes it easier or more difficult, respectively, to attain quotas. That is to say, thin materials make less kilos than thick materials, so a machine to make foil will not pass acceptance tests readily." In general, this meant that quantity reigned over quality—another throwback to the Romanian devotion to their five-year plan. Another point was that "if performance penalties are imposed, delays by Romanians will be attributed to the 'fault of the supplier.'" A third item was that although the completion deadline and performance criteria had been stipulated by both the buyer and the seller, the buyer was the one who actually stipulated whether the completion and performance tests met the requirements. But the requirements were also interpreted by the buyer, and the seller would be paid when the buyer was ready.

During the contract's preparation, GT found itself in the position of weighing the quality of its tires against the fees being offered by Romania. This meant the company was in the difficult position of having to determine the maximum durability and life expectancy of its product at a time far in advance of its actual manufacture. To underestimate the durability of its product would shortchange GT in terms of payment, but to overestimate the durability of the tires could prohibit GT from passing the acceptance tests or being compensated.

This situation arose when Romania laid out the standards that the tires were to meet and described the tests to meet those standards. GT felt that it should have been the one to lay out the standards, since the tests reflected its tire technology. Radial tire technology was part of the most current line, so extensive analysis over the years had not yet been possible. Thus, the company was dealing with technology that was not as familiar as the older tire lines. Furthermore, the negotiators on the project were representatives of the company; not all were from the engineering staff or had direct knowledge of the capabilities of the tires.

The quality of the tires produced also depended on the quality of the raw materials used in the production of those tires. GT provided specifications for raw materials, demanding that all tires carrying its trademark be manufactured from materials meeting certain standards. However, GT had no control over tire production if the Romanians chose to manufacture tires from materials that had not been continuously quality-checked as long as the Romanians did not use the GT trademark or indicate GT involvement. Checking raw materials from local sources in Eastern Europe was an expensive procedure, since the quality must be initially studied and then continuously policed to ensure consistently high standards.

General Tire was caught in a dilemma. Its immediate goals were to complete the construction and equipping of the tire factory in Romania and to run the acceptance tests on the tires being produced—in order to proceed to stage two of the agreement, the technical service contract. Production had to be started before the acceptance tests could be run so that the workers could be instructed in the operation of the equipment and processes. Since local materials were not as expensive as the standardized materials imported from abroad, local raw materials were used in early production runs. The tires that were produced, although unable to carry an indication of GT's collaboration, were still serviceable. GT had no control over this production, since the plant and the tires were the property of the Romanians. However, because the equipment was being operated, normal wear was also making the upcoming acceptance tests more difficult to pass. In the midst of this dilemma, the earthquake occurred and further delayed the tests. Shortly after damage to the plant had been repaired, the road test vehicle was destroyed, delaying the acceptance tests for another year.

Financial Flows

As a result of these delays, the additional time and effort brought additional expenses to General Tire without any supplementary payments. The best approach was to attempt to have the acceptance tests completed as soon as possible so that the scheduled payments would be forthcoming. The equipment that was repurchased after the earthquake was more expensive due to the passage of time and inflation. Since most of that equipment was purchased from other suppliers, GT had no control over the initial or subsequent purchase prices. Thus, GT's anticipated price list was, by that time, obsolete. In some instances, commissions were substantially reduced or negated. Equipment had been purchased under warranty, but delays and extended time schedules meant expired warranties. Any expenses incurred on the equipment after that period were then the responsibility of General Tire. One official pointed out that all additional expenses had to come out of the $1.5 million reimbursement, which was the total fee for the initial technology and machine installation.

Beyond the incidents already mentioned, a double taxation treaty went into effect between the United States and Romania in 1975. This meant that GT was then taxed by both countries, not just one or the other. These, too, were unanticipated expenditures.

On the other hand, it was to Romania's advantage to extend tire production time prior to the acceptance tests. The longer Romania could operate the plant before the acceptance tests the more tires could be produced and subsequently sold without any additional financial outflow. Furthermore, since the technical service contract would supply improvements in radial tire production for ten years, a longer period prior to its instigation provided for more possible developments and improvements that could then be applied to the Romanian operation.

GT noted that its contract contained one of the harsher penalty clauses, although, as yet, no penalties had been assessed and it did not anticipate that any would be applied, especially since the acceptance tests were so near completion. The contract applied a penalty based on a percentage of all equipment costs if certain pieces of equipment had not been installed by a certain time. Penalty clauses are often based on a percentage cost of just the undelivered equipment items—not on all machines. GT felt that the company should not agree to any more contracts containing such a penalty clause. If such a clause were mandatory, GT would then expect the prime contractor to take that responsibility. As a procuring agent, GT would take responsibility for any costs resulting directly from a loss of production due to events under its control or would agree to pay a certain amount for every day of delay of its making.

A counterbalancing aspect to the penalty clauses was the fact that the contract had no holdback clause. As noted earlier, many contracts within the socialist sphere require that a certain percentage of the initial fees not be paid until all work or purchases have been completed. If anything does not meet with total approval, that portion of the fee that was held back is in part or totally assessed as a penalty against the party that did not meet its obligation in full. The usual holdback percentage is 10 percent. It has been noted in connection with the Soviet Union that this 10 percent is usually never forthcoming, as some minor aspects usually are judged incomplete and the fee is withheld.

Sales of the tires produced in Romania were providing more income to pay the upcoming GT technology update fees. Production capacities in Romania are a state secret. GT officials did not even know the current tire production of the plant that they had provided. They were aware that improvements had been made in the plant's machines and that additional equipment had been purchased for the other tire plants in the country. They felt the production of tires in Romania could be exceeding even their expectations. However, regardless of the output, GT's annual fee would still be $500,000. The only way this amount could vary would be if another process for a different type of product were also purchased from GT; but this would be implemented only after a separate contract had been negotiated.

There is always the possibility that Romania will refuse to go into the technical service contract portion of the agreement, but it is hoped that this problem will not arise. General Tire felt that it had several points in its favor. First was the fact that the technology transferred to Romania was the state of radial tire development back in 1973-74. GT had made a number of improvements since that time, but these had not been transferred to Romania because the updating aspect had not been initiated. Romania would actually gain four to six years of additional technological developments prior to its ten-year contract yet would only have to pay for the ten-year period.

The aspect of inflation was a second point; the true dollar amount of the fee had dwindled substantially since the amount was first specified back in 1973. Thus, the cost to Romania would not be as great as initially laid out.

Legal aspects provided a third rationale. Because the entire agreement was a legal document, Romania's standing worldwide would be jeopardized should the technical service portion of the agreement not be honored. The Eastern bloc, as a whole, has the reputation for honoring all contracts, and this type of action would certainly be contrary to that tradition. On the other hand, Hayden reported that one middle-level State Department official noted that Romania was in default at the end of 1974 to "half a dozen" U.S. firms.[59]

The question arises as to whether General Tire would negotiate a similar type of agreement with regard to the divided fees portion with an update involvement of relatively long duration. In retrospect, it certainly would have been less complicated just to have built a tire plant and then to have left, much as Pirelli had done in Romania in the past. However, GT still feels that the transfer of its technology was not something that could have been successfully accomplished in a "one-shot" operation. This was due to the relatively long period of time that was required to set up such an operation. Improvements were constantly being made, and if these improvements were not implemented the tire plant was outdated from the moment its first total production was set into operation. GT saw itself in the business of long-term transfers and felt that the initial period was the most difficult. Relations between the partners usually became steadily better throughout the entire period of the agreement. Furthermore, General Tire saw the one-shot approach as a type of operation that literally created competition for a company, because little rapport or teamwork accompanied such a transfer. Through its extended operation, GT felt that it created additional supply sources and broadened its own base of operations.

For the Romanians, the cost of the project was listed at $30 million in loans for the equipment, including the $1.5 million paid to GT for equipment installation and initial technology, plus the additional expenses that Romania incurred in the construction of the plant. All payments for these initial purchases were received on time. The technical service contract fees were to be paid on a semiannual basis, with the first payment due six months after the completion of the acceptance tests. Should the acceptance tests not be successful, General Tire and Romania would have to reach another understanding. Failure to pass the acceptance tests would mean that Romania had not arrived at a "peaceful possession of the factory" and, thus, would have no right to GT's technology.

During the negotiation process, GT reduced its original statement of its technical fees, but the company felt that this reduction was proportionate with what would have been offered to other potential buyers. The fees would not have been out of line under the anticipated time schedule, but the lack of an escalator clause combined with the other unexpected delays made the financial situation much tighter than should have been the case.

There was another element in the financial flows of the Romanian project that involved mineral wealth. Romania, as an oil-rich nation, was also a source of petroleum-based materials. GT had hoped to acquire some of those materials but Romania would not sell them to General Tire at the price the company offered during the oil shortage in the mid-1970s. It should not be interpreted that Romania did not want to sell the materials to GT, but rather, as Hayden pointed out, Romania used its natural mineral wealth as leverage much as the more developed countries use their technology—by selling it to the highest bidder.[60]

It ought to be noted here that General Tire International was the company involved with the tire plant in Romania, but this is only one of the companies within the General Tire Corporation. As such, General Tire International was only indirectly connected to the other areas within the corporation that were involved in materials purchases. One GT official observed that GT subsidiaries do their own purchasing, and General Tire International only approves certain purchases. Thus, other affiliates of General Tire International had to become involved if certain purchases were desired from Romania. It must also be stressed that while Romania did possess some valuable raw materials, obtaining a particular quantity of a certain quality by a specific deadline required guarantees that such a supply would be available. Many suppliers often choose to buy from a source that, although more expensive, has been able to provide specified amounts according to a certain timetable.

Personnel

Training Assessment

GT's anticipated staffing requirements for the Romanian project were underestimated if only because of the unexpected occurrences and the extended time period required to complete the acceptance tests. The company estimated that 12 to 14 people would be involved full time for two to three years—or, in terms of man-months, involvement would range from 288 man-months to 504 months. Actually, the number of individuals from GT was substantially higher than expected, but the period of time that these individuals spent on the project was less than anticipated, so the final result was within the anticipated man-months projection. Approximately 40 of GT's top technical experts spent about 20 man-hours each on site in Romania, for a total of 800 man-hours. Combined with other GT personnel, the total came to approximately 300 man-months. Only one person in a supervisory position, GT's chief site representative, was still living in Romania at the time of the interview. Several others would be involved, however, until the acceptance tests were completed.

At one point, there were 67 individuals on site in Romania, including technicians from the various equipment manufacturers. GT has two families residing in Romania for three years each, including the family of Oz Perine, the project engineer who had the on-site responsibility for supervising the construction and equipping of the plant.

General Tire found that there was a bit of reluctance on the part of its employees toward living in Romania, although conditions there were relatively palatable compared with some other areas, like Saudi Arabia. Some of the problems encountered on a day-to-day basis were typical of those found in many developing countries, such as difficulty with transportation (especially private cars), a lower quality of housing, and the additional expense of imported goods. High quality food was not expected to be a problem in Romania, but the expatriates found that, to aid the hard currency supply, Romanians exported their best foodstuffs. The Romanian government's position was that the people had to work together, had to suffer a bit longer, so that factories could be built and the country could become prosperous. Ironically, the United States at that point was purchasing 100 percent of its pork requirements from Romania. However, the expatriates living in Romania temporarily had to survive on the local fare.

The agreement between GT and Romania called for adequate transportation and adequate housing for the expatriates, but it was soon discovered that each side had different definitions of adequate. For example, adequate transportation to and from the plant site was, to the Romanians, the local bus network. However, the expatriates soon found that this bus ran intermittently, at best, and not at all on weekends. The transportation problem was alleviated for the most part when GT bought an automobile to transport personnel to and from the plant. Housing, GT officials found out, should have been specified by the type of apartment or hotel rooms desired, right down to room size and toilet facilities.

The time spent on personal incidents was underestimated. For example, when a subcontractor had a heart attack, half the people at the construction site took the day off. Little things such as getting or renewing a driver's license took much longer than would normally be the case in the United States. Romanian holidays and national events also intruded on the project. The factory was located 50 miles north of Bucharest, along the airport road. If there was a demonstration at the airport, or if an important person was arriving or departing from the airport, people would be lined up along the road with banners or flags and all other traffic was stopped. There were times when the expatriates could not even reach the plant.

In addition to local disruptions or complications, however, the expatriates also encountered reminders that they were living in a controlled, socialist state. Their movements were constantly checked and their credentials examined even when they traveled from town to town. The expatriates would walk into their Romanian accommodations to discover that the contents had been examined in their absence, with pictures rearranged on the walls and other changes made. Harassment continued on the roads, where the expatriates received more than 30 traffic violations for such offenses as having dirty license plates. Once the expatriates were run off the road by Romanian party members in a black Mercedes. The Romanian government took the golf course away—stating that it seemed to be used only by foreigners—and made it into part of the adjoining Peoples Park.

On the other hand, the expatriates who had accepted the Romanian assignment had done so at least partially aware of the problems and difficulties they would be facing. All reported that their assignment was certainly financially beneficial. It did appear that no difficulties were encountered that were not eventually overcome.

The social isolation of the expatriates in Romania concerned the Americans more than the physical accommodations. The foreigners were really left as an island unto themselves. The families that lived there for several years noted that they were never once invited to a Romanian home for dinner, and they never ate out except at company dinners. They were also reluctant to invite any Romanians to their homes for fear of placing the individuals in a risky security position. The expatriates commented that there were two classes of Romanians: those who could associate with the GT employees and those who could not. It was understood that those who were able to have more extensive contact with the expatriates were those party members whose integrity was not questioned, or those who did not constitute a security risk should they escape to the United States. As such, the expatriates found that it was a society where all were equal, but some were more equal than others.

Despite the social and political restrictions, those who did have contact with the Romanians felt they were warm, friendly people. Three of the men sent there by GT married Romanian women, and all brides have since been permitted to enter the United States.

Security problems did not end on Romanian soil: GT officials noted that security was an important issue among those selected to come to the United States for training. Again, it was felt that several of those who came were sent in their capacity as party members, either to oversee the group or as a reward for services already performed. However, nothing was ever said in this regard. Those who appeared to enjoy life in the United States too much apparently jeopardized their futures in Romania, as they were treated as "contaminated" once back home.

Romanian distrust for the U.S. capitalists was apparent at the factory as well. In fact, Pirelli tire employees (or those employed at that operation) were used to check out the GT plant.

The expatriates believed that the traits exhibited by the Romanian workers were not so much the result of an undercurrent of distrust as they were indicative of the apathy that seemed prevalent throughout the country. Some of the expatriates involved in training the Romanian workers to operate the plant recalled their frustrations involving the bureaucratic type of behavior, a lack of tooling and machining skills in their trainees, improper maintenance of equipment, poor quality control and supervisory procedures at the management level, and, most of all, the lack of motivation in the workers. These were not impossible barriers to the project, however. General Tire International Project Manager Perrine described the Romanians as "good people." He noted that the Romanians had been cautious of him but added that the "cooperation was there." He realized that the Romanians had taken on an enormous project. In a training sense, he found that the Romanians had higher skill levels than many of their U.S. equivalents.

Perrine pointed out that a large amount of unlearning had been necessary because those involved in the GT factory had been formerly trained in the Pirelli and Dunlop facilities. On the other hand, he observed that his advice was not always followed because he "was a capitalist." Perrine felt that it was sometimes easier to introduce a new industry into a country than it was to update an existing one. When he had supervised the construction of a GT plant in Tanzania, the Africans had no knowledge of what had to be done, so instructions were always followed.

Perrine found that one had to operate within the Romanian system. For example, because bartering was a way of life, one bartered for the items that were needed, rather than just demanding them. He also found that he had no authority in the construction area. Once he had watched a foundation formed in the wrong location, but had no authority to stop the progress. Because Perrine and other expatriates were not allowed on the construction site during regular working hours, Perrine had to inspect the operation over the lunch hour.

GT felt that worker incentive was indeed lacking. This was apparently due to the fact that if mistakes were made, punishment was forthcoming. Workers did not seem to care as long as they did not have to make any decisions that could lead to mistakes. Their communication approach was to demand, not request.

The Americans also had a problem with petty theft. One GT official summed it up this way: "The people were relatively honest. If they had a chance to get something honestly, they would buy it—they wouldn't steal it. But anything Western that they could not buy—they stole. We were told 'there is no stealing in this country,' but after the equipment boxes were opened up, things disappeared. One example was the little colored plastic discs on some of the machines. These discs disappeared, but reappeared as bicycle reflectors. Another example might be windshield wipers. We had to take ours off each night to preserve them. They do not manufacture spare parts there, so there were no extra wipers to buy."

Problems occurred, too, in staffing the plant with Romanian workers. More workers were used than would be the case in the United States, including many more women. This tended to cause problems because many of the machines were scaled to the size of an average male. The women were far smaller, so many of the controls were out of reach for them. Thus, equipment was both overstaffed and underutilized at the same time.

Protection

Leverage Assessment

Domestic Company. Perhaps the most viable form of leverage available to General Tire at the present time is Romania's desire for GT's off-the-road tire technology. To have any credibility with General Tire (or with other tire firms for that matter), Romania will have to honor its present agreement.

It is interesting to note that a common item of leverage, the trademark, has not affected the operation in Romania, nor has the promised use of its logo provided GT with any apparent advantages. GT will not permit its logo to be used until the acceptance tests have been completed at the plant, but the Romanians are legally allowed to use the statement "made in collaboration with General Tire" on all tires produced at the plant and intended for use in Romania. However, the Romanians apparently felt that the production of tires as rapidly as possible was preferable to the use of any trademark or notation, and they have not indicated any GT involvement on any tires so far.

It would appear that the demand for radial tires in Romania was such that a particular trademark was of less consequence than actually obtaining the tires. It may also be true that GT's involvement with tire manufacturing in Romania was common knowledge, so that an indication of the tire technology's source was not necessary. Regardless, the trademark—an element of leverage used by other firms in dealing with Eastern Europe—was not sufficient in this case to overcome the other reasons prolonging the completion of the acceptance tests.

Host. GT's immediate aim was to progress beyond the acceptance test stage into phase two of the agreement. However, Hayden noted that there were at least three arguments against Romania's continued involvement with GT following the acceptance of the plant.[61] The first was that Romania was experiencing a hard currency shortage, partly due to technology purchases from Western nations. Thus, paying GT its annual technical services contract fee would be undesirable. The second was that tire technology is "fairly stable," with infrequent drastic changes. The basic process often does not change; in such cases updating may not be particularly crucial. The third aspect was that Romania had a reputation of not honoring all of its contractual obligations. GT had already experienced this type of behavior when the company earlier offered to help Romania dispose of some of its raw materials to help relieve the country's hard currency shortage; the offer had included GT as the first-choice purchaser. However, during the oil shortage when GT tried to obtain first bargaining rights to some of those raw materials, such as polyvinyl chloride, carbon black, and synthetic rubber, Romania sold the items on the Western European spot market where prices were 30 percent higher than those offered by GT.[62] The fact that the Romanians have chosen not to indicate GT's involvement in the tires currently being produced may be an indication of their desire for no future involvement as well.

It was noted that GT had the option of revising the agreement from a nonequity basis to one involving equity. Part of the justification for such a shift would have been to increase GT's leverage with Romania. However, leverage in the form of an equity basis appeared to have some major disadvantages as well as advantages. The deterrents did not center on nationalization, although that is a possibility whenever a company has an equity position in another country. Instead, they were based on the differing objectives of the two partners in this case. For example, Romania's objectives for a joint venture included access to hard currency and entry into Western markets. The U.S. company's objectives included access to Romania in particular and Eastern Europe in general. Thus, two different territories were involved. There was also a problem in the exchange rates between the U.S. dollar and the Romanian leu, since the current rate of exchange increased the cost of production in Romania and reduced the competitive aspects of such a joint venture in the world market. Thus, an equity arrangement did not appear to provide adequate leverage to offset the disadvantages.

Utilized Inspections

The upcoming acceptance tests will provide an opportunity for a major inspection, which will be the official one in terms of analyzing tire content and manufacturing procedures. Prior inspections had been utilized by various GT representatives to determine the proper construction and functioning of the equipment.

Patent/Trademark Defenses

Since no use had yet been made of the trademark or GT involvement, there had been no need to apply any trademark defense measures. No patent problems had arisen.

Legal Provisions Involved

The earthquake of March 4, 1977, was the only event for which the force majeure clause of the contract was invoked. Application of this clause indicated that the unanticipated delays in the project were not the fault of either General Tire or Romania, and that no penalty clauses or other actions were to be taken against the contractual parties as a result of the delays. The earthquake expenses were semireimbursed to GT by Romania (approximately 50 percent), since Romania paid all local expenses and a proportion of the remaining ones.

Hindsight

Factors That Made the Project Easier than Anticipated

General Tire's extensive experience in technology transfers of this type prompted the company to believe that it was not going to encounter any difficulties that it would not be able to overcome. Expectations were that problems would come up, but that nothing would be unsolvable. That still appears to be the consensus at GT, although the project has been more involved than anticipated.

GT did encounter several aspects of the transfer of technology to Romania that made the project easier than anticipated. The major discovery was that the Romanian's knowledge of the tire industry was much higher than expected. Furthermore, individual skill levels among the workers were also much higher, with the result that the training sessions were easier and the Romanians were able to grasp the essentials of radial tire manufacture more quickly. Because of the high skill level, it was also possible that some improvements could be made by the Romanians that would result in a reverse technology flow—that is, to General Tire from Romania.

In a sense, some of the greatest problems that GT faced—such as the damage incurred by the earthquake and the accidental destruction of the vehicle road testing the tires—might also be seen as beneficial. The disaster, which certainly affected both the Romanians and the Americans, had the effect of pulling both parties together in an attempt to repair the damages. Any time frame complaints that may have been instituted by either party were negated at the point of the quake; thus, no penalties were assessed. The road-test vehicle's accident certainly extended the time before the acceptance tests, but it also provided more opportunity to perfect the tires being tested, which may have helped the acceptance tests. Thus, even some of the greatest problems faced by General Tire in Romania may also be viewed as opportunities.

Factors That Made the Project More Difficult than Anticipated

Difficulties did occur. Some of those difficulties were expected, at least in part, but their full extent had not been anticipated. Some other problems, such as the earthquake or the road-test vehicle destruction, were not expected.

The underestimated problems included the lack of cooperation found throughout the Romanian system and the amount of red tape involved in accomplishing all tasks, large or small. The need for close administration, with the accompanying decision making necessary to maintain production, was also included in this area. Low worker incentives and morale were expected, but the lack of pride or care in the quality of the work produced and the lack of cleanliness were still surprising.

Progress of Future Expectations

Despite all the difficulties encountered, General Tire was interested in continuing its involvement in Romania for several reasons. First, since the company had made selling technology as much a part of its business policy as selling tires, and since it had equipped its operations with this in mind, it was generally interested in locating future technology sales. This included the area of off-the-road tires. Second, since the Romanian project required such close involvement between the Romanians and GT, this familiarity would make future projects with that country much easier.

GT felt that it would certainly revise its current contract format before entering into another agreement of the type now operating in Romania. The company was following its own advice in negotiating a three-way contract between East Germany, a French equipment supplier, and itself. GT would only be the equipment supplier, not the prime contractor, in this arrangement. In addition to purchasing the equipment listed by GT, the French company would also be responsible for its installation and the plant's construction. There was an additional advantage in the fact that the French company had done business in East Germany prior to this project and was familiar with the people and culture. Financing arrangements would also be more easily arranged, since the project could be financed through France as well.

The equity option on the Romanian project had not been acted upon during the first five years of the agreement and was considered an unlikely occurrence at this stage of the project. General Tire had concluded that the advantages obtained in switching from its nonequity basis would not be substantial enough in terms of additional operating leverage or profits.

GT was finding that its involvement in Romania had opened several other doors within the Eastern bloc and also with neighboring countries outside that specific region. The upcoming negotiations with East Germany have already been mentioned. The GT corporate staff entertained a delegation from the People's Republic of China (PRC) in January 1978. One of the company's greatest surprises was the fact that China was extremely impressed with GT quality because of the project in Romania. This familiarity with GT tires was in spite of the fact that the acceptance tests had not been completed and none of the Romanian-produced tires carried the GT logo or trademark.

Several other countries in addition to the PRC had expressed an interest in GT technology. Yugoslavia had approached the company with the possibility of establishing a tire facility in that country, and other inquiries had come from Poland, Hungary, and Czechoslovakia. Talks with the Soviet Union had already passed through several stages. Thus, it was felt that GT would be involved with the Eastern bloc countries in the future. In several respects, the project in Romania had provided a wealth of experiences that would be of great help on these upcoming potential projects.

Clark Equipment Company

Description Variations

Clark Equipment's project in Romania has apparently flowed more smoothly, on the whole, than its project in Poland. Clark officials felt this might be due to several reasons. One was that Clark had had two to three years' experience in transferring technology to an Eastern bloc country prior to the start of the Romanian project. In addition, the Romanian project took longer to negotiate, which gave all parties an opportunity to become personally acquainted prior to the start-up of the operation. Clark discovered that U.S. companies, in general, conclude arrangements on the basis of the company's name being known—whereas the Eastern European approach concentrates on the <u>people</u> involved with the negotiations. They must become acquainted with their counterparts before the negotiations are accepted and concluded.

Clark felt that its representatives had an extremely good relationship with Romania, right from the very first day. Part of this was perhaps due to the continuity that Clark encountered in dealing with Romania. The same individuals tended to stay in their positions, and Clark was able to establish a basis for stability that had been lacking in Poland, where personnel and positions tended to be shifted more often.

The initial contract between Clark and Romania, represented by all three agreements, involved transmissions (but not for crawler tractors). The amendment that provided for those transmissions might be considered a separate contract. The transmissions were created through a joint development process wherein Romania made all the components, paid all the costs incurred by Clark in testing out the transmissions, and will also pay Clark a fee for all transmissions of this type that Clark is able to sell outside Romania. This transaction went somewhat beyond the area of licensing. Romania needed a transmission for a crawler and wanted it in metric scaling. By working jointly on this need, Romania has the crawler, and Clark has started its metric changeover. They made the drawings in the United States and the parts in Romania. Both sides independently tested the product and then compared results.

The amendment that was added made changes in several of the agreements. In the Technical Assistance License and Cooperation Agreement (TACA), changes dealt with the additional products covered under the agreements, the engineering information to be furnished by CEAG to UT, CEAG's nonavailability to enter into other similar agreements for products covered by this agreement, CEAG's aid to UT involving the extent of the information to be transferred, and changes made in the scheduling of payments of the disclosure fees.

The amendment, signed April 10, 1975, also noted several changes in the marketing agreement between CIMSA and UT. The first change added two crawler tractor transmission models to the information contained in the technology transfer. These were models LT7758 and LT7759: a 180-horsepower transmission and mating torque converter, and a 120/140-horsepower transmission and mating torque converter, respectively. The models, including those added by the amendment, included model series numbers 2000, 3000, 4000, 5000, 8000, and 16000.

Unlike the problems that arose during the joint development project between Clark and Bumar, the joint development project that came about between Clark and Universal Tractor appeared to flow smoothly. Romania wanted 20 months for its responsibilities. The final outcome required 30 months, as CEAG and Romania built a prototype of a transmission for a crawler tractor. Previously, the transmission for this bulldozer had been imported from West Germany. Romania, however, wanted to become more self-sufficient in this regard, as imports were particularly undesirable in light of its hard currency shortage. The transmission had to be fully metric in scaling. As stated in the agreement, all products developed by Romania or by both partners related to the agreement were to become the property of both partners. Clark now has the rights to this metric transmission for its own use should it desire this item.

However, Clark's arrangement with Romania, for instance, is completely different. Clark has no equity in the Romanian project. Clark's personnel worked with Romania's in creating this product, but the product was designed by Romanians and was paid for by Romania.

One unit of the 180-horsepower crawler transmission, which had been prototyped at Hidromecanica late in 1977, had been sent to the Clark Equipment plant in Jackson, Michigan, for inspection and tests in 1978; three additional prototypes were being machine-tested at the same time in Romania. Romanian production schedules for this transmission model were slated for the second half of 1978. Additional models scheduled for the 1978 production in Romania were the 4000 series power-shift transmission, the C5000 series torque converter, and the 140-horsepower crawler transmission.

Current Corporate Perception of Benefits from Agreement

Clark's expectations are apparently materializing in Romania, although at a slower rate than originally scheduled. The facility—as a source of transmissions or profit center for license fees and royalties—has been slower in developing production levels than time schedules had first indicated. However, offsetting this deficit was the increase in component sales and exports to Romania to facilitate its production.

An additional benefit, which was anticipated as a possibility—though not to the scale that has materialized—was the joint development project involving the metric crawler transmissions. Clark had not originally foreseen the additional fees that could result from such an endeavor; it will also have rights to the use and production of the units when fully developed.

Negotiations

Clark's experience with the government in Romania during this project has been extremely cordial. Romania received most-favored-nation status prior to the agreement, which was a great help in terms of financing and arranging the project, but Clark felt that the cooperative nature of the officials seemed to run far beyond normal political relations.

Contract Format Revisions

Clark felt that its contract with Romania was most beneficial. Based on the success of the project thus far, the company's analysis of some of the negotiation arrangements, negotiators, and the contract format might be helpful for others thinking of becoming involved in a similar venture.

Levels of Negotiations. Clark found that there was no need to organize meetings with government officials—it was necessary to wait for an invitation from such an individual. An invitation was usually forthcoming only after the initial arrangements for the technology transfer had been made with the appropriate foreign trade organization. That unit tended to bring all individuals together, including ministers and partners from the companies directly involved with the transfer. The ministers have the ultimate approval power.

For corporations based in the United States, the starting point is not directly in Eastern Europe but usually with the Bureau of East-West Trade in Washington D.C. This bureau will usually contact representatives in Eastern Europe after a preliminary analysis has approved the technology for transfer.

Negotiators. The representative or representatives from the U.S.-based company (or any other company, for that matter) need not be lawyers or technicians, but Clark has found that an individual having familiarity in both of these areas can be invaluable. Such an individual would be able to negotiate from both angles and often facilitates the negotiations. Selection of this individual as the official corporate representative is apparently one of the best ways to successfully negotiate a contract, but Clark further admonishes that, in addition to the aforementioned skills, this negotiator (or negotiators) must still expect lengthy and drawn out discussions and have the patience to endure them.

Clark has found throughout its dealings with Eastern European countries that negotiations require far more personal acquaintanceship than is usually the case in the United States. The time spent in achieving mutual rapport over the negotiation table is actually a means for facilitating the project once the agreements have been concluded.

Clark also admonishes the negotiator to be prepared for lengthy sessions, and to expect the negotiations to persist over a long period of time. Regardless of the time required or needed, however, Clark also points out that no contract should be signed unless it is totally satisfactory, as mistakes or items overlooked can be very expensive or detrimental.

Contract Issues. Clark points out that all contracts with Eastern Europe should be exceedingly detailed, because this contract will be followed to the letter—regardless of the actual stipulations that were accepted. Great care should be used in scrutinizing each aspect contained in the contract prior to the final signing. In the United States a contract's wording is sometimes taken to court for the final interpretation; in the Eastern bloc, the contract itself is often taken to be the final arbitrator. Thus, such details as financial considerations should be clearly laid out; Clark also recommends that a final ceiling be placed on all financial commitments.

Contracts should contain a "two-way street even though the traffic in one direction may be considerably greater than that in the other." Examples of such clauses include governmental approvals (both sides), nondisclosure of technology of military importance, two-way development flows, and inventions created from that immediate technology.[63]

Clark tries to include clauses that leave open the possibility of joint development, as this type of activity tends to bring the partners closer together and often results in new variations if not totally new products. One common possibility for such an activity, Clark points out, is the conversion of U.S. products to the metric system.

Despite the haggling that characterizes the negotiation sessions, Clark has found that the atmosphere changes greatly once the contract is signed. The process of getting the final approval might be frustrating, but once this has been obtained, the partners (at least in Clark's experience) uphold the agreement to the letter.

Time Frame

Tooling and machinery procurement for the transmission plant in Brasov was progressing favorably as scheduled, as of 1977, but was not completed by that year.

Because it was supervising the building of an entirely new plant in Romania, from scratch, Clark had expected production to begin in 1977, or approximately two and a half years after the start of the building. But production of the transmission was under way, officially, three and a half years, or by mid-1978, a year behind schedule. However, given the enormity of the project, Clark felt this delay was not too bad.

Financial Flows

Despite the smaller size of the operation in Romania, Clark was surprised to find that, from the start of the Romanian project in 1975 through 1978, the net cash flow to Clark was twice the size of the net cash flow to Clark from Poland, despite the fact that the Polish operation had been under way since 1972. This included both fees and sales for the entire extent of each respective project from its official starting date.

Clark shipped approximately $2,400,000 in axles, transmissions, and torque converters in 1977 for orders entered by Universal Tractor in 1976. Greater levels of shipments were expected in 1978, but some of the bureaucratic problems inherent in the Romanian foreign trade system had to be overcome before this could materialize. In addition to the initial disclosure fee, Clark received an additional payment for its cooperative help on the joint project.

In retrospect, the sale of component parts to Romania provided Clark with an even greater benefit than had been originally anticipated. Romania wanted to create a construction-making industry, of which the production of transmissions capable of driving construction vehicles was only one step. Bumar in Poland, on the other hand, represented a more mature industry interested in supplying its own needs. Component sales to Romania have totaled over $1 million annually. This represents a total net gain, Clark pointed out, because the company was not in the market at all prior to the Romanian agreement. The initial products contained 10 percent locally produced components; the other 90 percent was composed of imported components, the majority from Clark. However, Clark realized that these component sales would drop off sharply as locally produced components of the axles and transmissions expanded.

As with negotiations, Clark realized that products often will not sell on the basis of the company name alone, but only after the individuals representing the company become known. The slump in sales that hit Eastern Europe in the latter part of 1974 and persisted through the start of 1977 did not have as much impact on Clark's Romanian operation as it did on its venture in Poland—for reasons other than Clark's reputation, however. The lessened impact in Romania was apparently due to the fact that the Romanian project had begun later and, hence, was in the building and equipping stage, not the manufacturing one. The slump in sales did not hit Romania because the product was not there to sell in the first place. Since the Romanian agreement was signed in 1975, the project was barely under way during most of the economic slump. There was a slight drop in sales from 1975 to 1976, largely in terms of a percentage of total sales in Romania, but quite slight when compared with sales throughout Poland or the entire Eastern bloc. It was presumably due to the fact that some transmissions had been imported but—with UT's potential for manufacturing its own materials—their subsequent sales were postponed. This postponement would also have been encouraged by the recession, which lessened the availability of hard currency. In fact, from 1976 on, Romania's sales in transmissions rose dramatically, whereas sales in both of those product categories were falling in Poland and other areas. Clark reported that the license agreement with Romania led to substantial orders for component parts.

The Romanian project called for a much higher percentage of imported, U.S.-made components in the final assembly of transmissions than Clark's manufacturing contract in Poland. As of 1977, U.S. input in the final product (in Romania) was about 80 percent. It must be pointed out, though, that this was barely two years after the project had been started. Furthermore, the Romanians were not as prepared in terms of support industry as, for example, Poland, where the percentage of imported components in the final product was approximately 30 percent. In retrospect, the export market appears to have been more anticipated than what actually occurred up to 1977, which Clark hoped to find in Eastern Europe. However, Clark did feel that more mileage had been gained from its transmission, which might be interpreted as aiding the overall product development costs.

Other financial factors tended to counterbalance each other. Whereas the fees initially requested in the negotiations with Poland had been accepted without much discussion (and, hence, were higher than Clark had originally anticipated), the fees requested in the settlement with Romania had been compromised to the satisfaction of both parties. However, the joint development project with Romania appears to hold more promise than originally anticipated.

The lower license fees received from the Romanian operation, due to the slow start-up, were disappointing, but since the license fees were to run for ten years from the time the product was first manufactured and sold, these fees were expected to pick up over the time of the agreement.

Personnel

Training Assessment

To date, more than 1,000 individuals have been involved in the Clark project with Universal Tractor in Romania, including all workers in the Romanian factory.

Domestic Company. Over 80 Clark employees from around the world have participated in the technology transfer or training sessions connected with Romania. Clark reported no unusual difficulties or problems with its employees and found the climate within Romania accepting and responsive.

Host. Clark found that the officials or top workers in Romania were more motivated than the general factory workers because cash bonuses were available. Common motivational problems prevalent throughout Eastern Europe, such as quality control and worker indifference, were encountered, but Clark did not find this a deterrent to the project and noted that the Romanians were, on the whole, skillfully trained and anxious to have the project succeed.

Protection

Leverage Assessment

Domestic Company. Clark's initial control rested on the inspection and quality control standards that had to be met before the Clark trademark could be applied to the product. The marketing agreement was also a form of control because Clark will be selling items for UT, actually helping UT to export the product.

Host. The increased rapport between Clark and UT apparently reduced the usual emphasis on controls. Romania has met its financial commitments and yet also holds the promise of increased involvement with Clark as an enticement for the future.

Utilized Inspections

Quality control has been a concern of Clark's throughout the agreement, since its trademark is involved, but no inspection problems have come up.

Legal Provisions Invoked

One distinction that Clark found between Poland and Romania was that Romania would use intermediaries to help negotiate or settle disputes before an issue was sent to the arbitration board. Poland did not do this and, as a result, Clark has found it easier to work with Romania.

Hindsight

Factors That Made the Project Easier than Anticipated

Clark found that the Romanians were less familiar with the manufacturing and assembly of the transmissions and, therefore, tended to follow Clark's advice more closely. The Poles, who had extensive experience in similar (although not updated products), tended to do things their way until convinced to follow Clark's advice. Thus, Clark actually found it easier to work with the Romanians.

Factors That Made the Project More Difficult than Anticipated

The basic problem was the longer time required before the start-up of production. Clark felt that the problems encountered in Romania were really minor in scope and were the kinds of things that arise around the world during the construction of a new plant. Questions on the type of processes to use, for example, or adapting to local conditions in the most efficient way are illustrative.

Problems always arise when transferring working methods from one culture to another. As an example, Clark noted that the cement floor in the Romanian plant was not sealed, a procedure that had not been pointed out because it was so taken for granted in U.S. building construction and finishing. On the other hand, once such items are noticed, they can usually be remedied easily.

Clark found its involvement with Romania was not out of character with its knowledge of the constraints found throughout Eastern Europe. However, despite this general expectation, the controls for letting individuals out of Romania were so tight that it was difficult for individuals to obtain exit visas to come to the Clark training sessions in the United States. In this regard, Poland was relatively free in its dealings with the West. Romania also had a lower standard of living than Poland and an accompanying lower stage of development, with its attendant problems.

Progress of Future Expectations

No further arrangements materialized between Clark and Romania, but several other areas of interest were developing with other representatives in the Soviet bloc. Clark sent a proposed technical cooperation protocol to the State Committee for Science and Technology in late 1976, but the proposal was viewed as too broad, and Clark had been requested to detail specific areas of cooperation. This matter was also being considered by the U.S.-USSR Trade and Economic Council.

Clark had begun discussions with the USSR in 1976 regarding a lift truck license. These discussions continued throughout 1977, and a product and production study was presented to both Clark and USSR officials in 1978. Clark noted that the USSR had been evidencing interest in lease agreements intended to cover various types of products manufactured in the West and required by the Soviet Union.

The lease arrangement was being suggested as an alternative method of financing purchases without increasing hard currency debt. Clark prepared a lease agreement draft for presentation to Soviet officials in March 1978, paralleling a program supported by the U.S.-USSR Trade and Economic Council, which was attempting to open up additional trade between the two countries.

HUNGARY

International Harvester Company

Description Variations

The technology transfer to Hungary was enthusiastically received and the products are currently being manufactured there. The oldest of IH's multiple agreements was barely two years old at the time of this assessment, so this was a most positive sign predicting that IH would seek more ventures of a similar form in the future.

The 1976 tandem axle technology transfer, supplemented by a plows and harrows agreement in 1977 and a cycloplanter agreement in 1978, was joined by a grain drills agreement in 1979.

The results reported here concentrate on the variations that occurred between the initial expectations and the final outcomes. However, these details may not accurately portray the overall success of the transfers, since these results and conclusions are based on a minimum of time. Undoubtedly, additional changes will occur before the agreements conclude in the late 1980s.

Technology updates were involved because IH was also selling components to Hungary in various manufacturing and assembly ventures, so it had to ensure that all components were interchangeable. Generally speaking, all technology had been updated and was current on all projects in all locations throughout Hungary.

The axles manufactured in Hungary were up-to-date and, according to IH, certainly comparable to axles made by competitors in the United States. However, since the axles were of IH design, all service and replacement business came to IH. Furthermore, since these axles went into trucks, this provided the trucks with a higher percentage of IH parts. Hungary and IH were working on a joint development project in connection with these axles, but the end result was to refine and improve the existing axle rather than develop a new one.

Several compromises were made to allow for a smoother-running operation. For instance, the original plans called for 27 inspection points, but this was cut back to 15. IH did some marketing of the farm implements manufactured by RABA to help the Hungarians in selling their product. These implements were not to be sold in the United States but were designated for other areas in and around Europe.

Current Corporate Perception of Benefits from Agreement

The apparent opportunities afforded to IH and RABA via their axle agreement did not change to any great extent during the first three years of the project. IH saw the opportunities being afforded by the RABA agreement, at least in regard to the axle technology transfer, as relatively similar to its project in Poland. IH's primary reasons for the arrangement were to obtain a source of supply of axles for its use elsewhere, and to sell component parts to RABA that went into the construction of the axle and farm machinery. Actually, since few component parts were purchased for axle production and assembly, most IH component sales were in the farm machinery line. The axle arrangement was a form of subsidy by IH for the farm implements arrangement because IH had contracted to repurchase the axles. This gave RABA the requisite funds to purchase components for the manufacture and assembly of the farm implements.

All of its arrangements with Hungary increased IH's toehold on the Eastern European market. This gave IH direct access to the potential Hungarian market, while it reinforced the company's position as a product supplier throughout Eastern Europe.

The fees paid to IH for its technology and the subsequent royalty payments (based on sales of IH-sponsored products) were another reason for IH's desire for such arrangements, although this reason ranked third or fourth in priority. The fees were desired, certainly, but IH found that additional expenses in training often tended to offset if not negate these fees. However, the long-run potential of these projects provided IH with enough justification to continue to seek further involvement along similar lines in the future. RABA wanted to increase its export market both within Hungary and outside. Its goal was to become an integrated farm machinery supplier, and it needed access to a source of technology in order to obtain a full range of farm machinery—such as IH was able to supply. RABA obtained many of these goals through this transfer.

Negotiations

No major changes were reported in the contract format since the agreements had been signed. The rapport and progress thus far, in fact, had encouraged IH and RABA to seek out additional technology transfer areas.

Time Frame

On Schedule

Delays occurred, preventing the time schedule from correlating with earlier estimates.

Behind Schedule

The time frame for the Hungarian project suffered from delays. Two time problems occurred during the initial setup period, caused by equipment delays and tooling problems. The equipment that had been ordered and purchased for the project was late; hence, it could not be installed according to the original time schedule.

The second problem was that the product, when it did come off the production line, did not meet specifications. These time problems were not anticipated, as IH had accepted the time frame as laid out by RABA.

Financial Flows

Payments

Financial flows to IH from Hungary were behind schedule. These fees were related to the production schedule, however, not the disclosure fee payments.

Purchases from the Domestic Company

More components or pieces of farm equipment had been purchased by RABA than initially estimated, although fewer total units of farm implements had been sold to Hungary. Interestingly, two different organizations were purchasing whole units or components, but overall sales of the equipment had fallen throughout Hungary (a trend reflected throughout Eastern Europe). Entire units or whole pieces of farm machinery were purchased through the foreign trade organization operating out of Budapest. Components or other pieces were purchased by RABA for use in its manufacturing and assembly operations. The RABA purchases were higher in volume.

Personnel

Training Assessment

IH found that Hungary needed more technical assistance than anticipated and that a far greater training effort was required than originally realized. IH had previously encountered training difficulties requiring additional time in Poland and should have been prepared for similar experiences in Hungary. For some reason, however, IH felt that the Hungarian workers would not need such assistance. These training tasks, combined with the expense of travel back and forth, negated a major portion of the profit from the technology fee.

On the other hand, unlike Poland, the Hungarians were quite self-sufficient in several other aspects. For example, in Poland it had been felt that there was "a need to hold the Poles' hand," whereas the Hungarians were more capable of working on their own. Also, in Poland additional time had been required by executives for negotiations, but negotiations with Hungary tended to be less complicated or involved.

IH found that the transfer of the process portion of the technology to Hungary was considerably more difficult than anticipated, and a number of IH technicians were sent to Hungary to review the process as it was operating in the plant there. It was determined that because RABA had been manufacturing axles for years, the workers felt they knew how to make them—consequently, many IH instructions were ignored. This situation was even more acute than the one that occurred in Poland where the employees—although they had been involved with the manufacture of construction equipment in general—had never manufactured crawler tractors prior to IH's involvement. The problem appeared to lie with improper sequencing and with a lack of correct testing procedures, not in the use of the equipment per se.

One of the major training tasks in Hungary was the improvement of poor quality control procedures. Other problems such as a lack of workers' skills or inadequate worker incentives were not as evident in Hungary as in Poland.

Protection

No changes appeared to have occurred since the initial agreements had been formulated. If anything, the continuing rapport between IH and RABA would probably be of greater benefit in negotiating future agreements or in working out problems on the present transfer than even another form of leverage. On the other hand, part of this rapport may exist because the leverage was already there on both sides.

One situation with a potential for misunderstanding did develop problems, probably because some leverage did exist there. Fortunately, the predicament was recognized and corrected rapidly. It centered around use of the IH trademark, which is granted only after the products have passed all inspections and cleared all quality control checks. In the case of one product in Hungary, however, RABA felt that IH was deliberately slowing down the program, presumably because IH had agreed to buy back some of the products. RABA interpreted this slowdown to mean that IH did not want to buy back the products, and the company was using the trademark quality checks as an excuse. (If its trademark was not present, IH was not obligated to repurchase the goods.) When IH became aware of RABA's assumptions, the company assured the Hungarians that this was completely wrong. and that IH was only concerned that product quality be maintained at the corporation's worldwide high standards. In this regard, Poland had been exposed to Western quality control standards more than Hungary. This was an aspect of the Hungarian accord, not immediately apparent in the timing sense, that involved greater difficulty in Hungary than in Poland.

Hindsight

Factors That Made the Project Easier than Anticipated

Unfortunately, none of the elements appeared that were initially anticipated to make the Hungarian projects easier: less complexity, prior experience by RABA employees with axle production, and IH's past experience in the nonequity form of technology transfer. On the other hand, the rapport between RABA and IH was comfortable and delays in training and productions were relatively brief. The fact that production commenced within two years of the start of the agreements emphasizes the positive working of the project.

Factors That Made the Project More Difficult than Anticipated

The fact that IH was not anticipating many difficulties may have inadvertently made the project more difficult. For example, IH tended to accept RABA's time schedule estimate when it might have been more accurate had IH plotted its own, based on its own experiences. Had the extent of untraining been realized, more emphasis might have been applied in urging Hungarian workers to follow IH instructions. In general, prior experience certainly is more of an asset than a deficit. It also appears that the format of the agreement has an impact on the success of the project, and IH certainly benefited from its prior experiences in this regard. However, operating solely on past occurrences could distort one's perception of new situations.

Progress of Future Expectations

Additional projects are in the works, based on the growing success of the IH-RABA agreements. The current agreements between IH and RABA deal with truck axles and agricultural equipment, but the possibility of an exchange-type of program between RABA in Hungary and Bumar in Poland was being explored, with Hungarian products to be bartered for Polish-made construction equipment of IH design.

YUGOSLAVIA

The Bendix Corporation

Description Variations

Four years after the onset of the agreement (just after the production start-up), a comparison of expectations versus outcomes of the Bendix agreement with Vartek found relatively few items of disagreement. The price Bendix placed on its technology had been accepted as fair and equitable. Production had been delayed for almost a year, and the friction materials, when produced, had not been up to specified quality, but these problems had been overcome. Working relationships between Bendix and Vartek were excellent, to the point where Bendix was seriously considering exercising its option of joining Vartek in an equity position or joint venture.

One situation arose in connection with the workings of this agreement that illustrates both the advantages and the disadvantages of MNC involvements of this type. The broader range of services that an MNC can provide is usually an advantage, as in this instance where the disc brakes and the friction materials were able to be provided. Through this agreement, Bendix, in Michigan, was able to provide goods to Yugoslavia that the French subsidiary of Bendix, DBA, was unable to furnish. Because of its multinational operations, Bendix was able to provide a service through its wider range of technology and products that would have been impossible for just one company.

However, this multinational factor may also generate disadvantages, problems, or misunderstandings. One such incident occurred after the agreement with Vartek was under way. Bendix acquired another company after January 1975 that held a formula for another type of organic friction material. Vartek contended that this formula should be provided to the Yugoslav company, since it was an advanced form of the technology that, according to the contract, Bendix was to provide to Vartek until 1983. Vartek's position was that all technology related to the brake friction materials, which was available to Bendix both directly and through its subsidiaries, should be covered by this agreement. Bendix, on the other hand, felt that only the technology held by the corporation based in Southfield should be subject to this agreement. Its subsidiaries operate independently and information held by them should not be involved. In this instance, a compromise was reached. The new formula was given to Vartek, but the training sessions on the new formula, provided by Bendix, were conducted at a Bendix subsidiary in Spain, not at Troy, New York, where the other training had taken place. No additional payments were requested.

Current Corporate Perceptions of Benefits from Agreement

The reasons for creating the agreement have remained constant, but the emphasis has shifted. Initially, the foremost rationale was the monetary gain from the fees paid for the technology. However, the compatibility that Bendix and Vartek achieved helped to change the emphasis from immediate monetary gain to the establishment of a toehold in Eastern Europe, which could lead to greater involvement for Bendix both within Yugoslavia and throughout other countries in Eastern Europe as well as in the Soviet Union. At the same time, the agreement with Yugoslavia was becoming increasingly distinctive, primarily because of Yugoslav characteristics that distinguish that country from others in Eastern Europe. This was particularly true regarding its relative freedom from controls and accompanying lower emphasis on security.

Negotiations

Bendix felt that the agreement was concluded without unpredictable difficulties. Cognizant of the independent nature of companies in Yugoslavia, compared with most of the other Eastern European countries, the corporation still felt that the Vartek company would have to await approval from the government. However, it appeared that there was no lengthy or heavy government scrutinization. Neither was it requested that the agreement be scaled down, either with regard to its duration or the amount of the requested payments. Bendix arrived at the price placed on its technology after consideration of several factors: its importance to the buyer, its estimated remaining product life before replacements took over, the likelihood of suit or patent infringements, an analysis of potential competition, and the estimated production demand.

Time Frame

Behind Schedule

The start-up of actual production was behind the projected schedule by six months to one year; it finally began in full in early 1979, four years after the agreement was ratified. The delay was attributed to the inexperience of the Vartek company and its employees in production of this type. An accompanying problem involved the inadequately low quality of the friction materials that were being produced. However, additional specific training helped to overcome these problems so that current production met both quantity and quality specifications. This training specifically concerned the selection of raw materials and the qualifications or acceptance procedures needed to approve those raw materials. The delay did affect Bendix, since additional visits by its corporate personnel were required, and the payment schedule annually collected was limited to minimum royalties. The incentive to meet anticipated production quotas was high, however, because the sooner production was up to expectations, the sooner Vartek would be able to satisfy some of its contractual obligations.

Other Aspects—Extension

The projected time frame designated a ten-year limit, but it is now expected that an extension will be requested regarding technology updates. No official request has been made to exercise the equity option and, at this point, only six months remain in which to do so (by January 24, 1980). Since production has been in progress only six months using the original formula, no updating in the technology has been added, although it is highly possible this will be done before the agreement is concluded.

Financial Flows

Payments

Bendix noted that lump-sum payments of specific amounts are requested in contracts of this type within Eastern Europe more often than payments based strictly on sales or royalties. The agreement stipulated that Bendix was permitted access to Vartek records for verification of sales, although no problems were anticipated regarding the accuracy of these royalty payments. It was also pointed out that since the friction material was used in the production of automobiles a fairly accurate count would be possible based on the number of automobiles manufactured each year.

So far, only the minimum royalty fees have been paid to Bendix. Vartek has submitted a quarterly report of sales, documenting the amount of friction materials sold. Three annual payments have been received in addition to the initial disclosure fee. The annual fee payment schedule has been approved by both parties through quarterly reports on production and sales.

Eventually, the minimum royalty payments to Bendix should increase proportionately as more friction material is produced. So far, Bendix has not had any problems with the payments in U.S. dollars, although the set fee reflects the rate of inflation only through increases in the selling price of the product. No problem occurred in connection with the payments—regarding either form, timing, or collection.

Purchases from the Domestic Company

Vartek made no major purchases of completed organic friction materials. This was not expected, but it is always a possibility when a firm has a contract to fulfill and production is delayed. The anticipated two months' supply of nonassembled friction materials was purchased, but more as a training aid than to fill a certain quota. Vartek also imported resins from Bendix, U.S.A., as of September 1978 but this need was not expected to persist beyond a year.

Bendix found that Eastern European countries will not make large quality purchases, in most cases, and attributed this behavior to the shortage of hard currency in those countries, which made such purchases relatively difficult.

Purchases by the Domestic Company

Bendix made no purchases in Yugoslavia of any consequence in connection with this project.

Personnel

Training Assessment

The high morale of the Vartek employees was somewhat unusual in comparison to Eastern Europe as a whole, but the reasons for this apparently lie partially within the structure of Vartek itself, and partially with the governmental structure of the country. The company has had a very good reputation, at least in its textile production, so an accompanying factor of employee pride in craftsmanship already existed. The company also had a high level of community involvement. Thus, the delayed production schedule and the accompanying low quality output were not results of low employee morale and, in part, represented a challenge to the workers.

There had been a minimum of language difficulties up to the time of this interview, although several problems apparently resulted from the necessity for language translation. If confusion or misunderstandings occurred, a closer assessment of what was being said usually resolved the difficulty.

Bendix has a communication policy of keeping a written record of all interchanges. It is felt that this custom has prevented many communication problems from occurring or increasing.

At least five visits to the site in Yugoslavia were made by three different individuals from Bendix; each visit lasted approximately one week. Reciprocal visits were made by two Yugoslavs to the United States, again for a week each time.

Intensive training was started in 1977 and concluded, for the most part, that same year. At that point in time, training involved variations resulting from the revised formula; in the future, it will most likely be limited to updating or improvements.

Vartek paid for all travel expenses, as stipulated in the agreement. The Yugoslavs paid their expenses directly, and expenses incurred by Bendix technical and training representatives were invoiced to Vartek for reimbursement.

The delay in production start-up was attributed to the workers' inexperience with the manufacturing process, but this was solved by additional training directed toward specific trouble spots.

Other Personnel Involvement

There have been no security problems in the agreement either on the part of the United States or Yugoslavia. This technology was not viewed as a security issue, so no U.S. governmental approvals were necessary. The contact between the Yugoslavs and the Americans was open and unrestricted. However, it should be pointed out that no U.S. assistance was involved with the construction or equipping of the factory beyond the suggested list of equipment items. Furthermore, the training visits involved a very small number of Americans in Yugoslavia, and Yugoslavs in the United States, and they lasted for the relatively short period of one week in each instance—so contact between the two groups has not been highly extensive.

Protection

Leverage Assessment

Domestic Company. Vartek's incentive to start production as soon as possible was the fact that it had to pay Bendix a minimum royalty for each year, whether the product was being manufactured or not. Bendix maintained a high degree of leverage in that the company's expertise was seen as a source of knowledge from which to work through training problems and overcome any difficulties. Furthermore, most delays did not cost Bendix any additional expense as Vartek would be the party most concerned about getting the production moving.

Host. Bendix made its hopes of extended and increased involvement known and Vartek, as well as Yugoslavia in general, would be the contact. Therefore, there was great concern that this immediate project should be successful so that more would be forthcoming.

Utilized Inspections

Inspections were used to determine whether the quality of the materials was up to specifications.

Patent/Trademark Defense

No patent suits or infringements have been brought up, and Vartek has not requested any export permission for the materials, apart from their use as original equipment on cars. Use of the phrase "made under license by Bendix" beyond the termination of the contract will cease if no extensions are applied to the agreement.

Legal Provisions Invoked

Thus far, arbitration has not been invoked, nor has there been any provocation for considering arbitration. There have been no situations calling for a termination of the agreement due to default, and no instance has arisen that would point out the lack of a true force majeure clause in the contract, or force an interpretation of the contract on such a point.

Hindsight

Factors That Made the Project Easier than Anticipated

The rapport between Bendix and Vartek has undoubtedly been the greatest asset to this venture. Their mutual anticipation for the extension of this agreement, as well as for additional projects, has provided an impetus to make this agreement as successful as possible. The lack of unexpected or difficult problems has provided a relatively smooth period of operation throughout this agreement. The control that Vartek exerts over its responsibilities has also provided an adaptability that is sometimes lacking in some of the other Eastern European nations.

The relatively small size of the project made its execution easier. Bendix did not have any construction or even equipment supplier responsibilities but only functioned in an advisory capacity in this regard. Furthermore, unlike agreements that require the seller to cover all expenses within the established fee statement, this agreement permitted Bendix to be reimbursed for all actual expenses over and above the stated disclosure fee and subsequent royalty payments. This was a possibility because this project had a minimum of travel or training expenses compared to most other projects involving a high degree of technology transfer. Bendix also benefited in a general way from the presence of its subsidiary's agreement with Vartek, which was much greater in scope than the specific friction-materials production project analyzed here. The subsidiary, DBA, helped to smooth the way in many cases, setting a general form of operation that Bendix (U.S.A.) was able to follow as well.

Factors That Made the Project More Difficult than Anticipated

The only difficulty of any magnitude encountered by this point in time was the delay in production start-up time, with the accompanying low quality problem in the materials that were produced. This was attributed to the inexperience of the workers, who had a textile production background, and to the lack of suitable raw materials. A lengthy acceptance procedure had to be instituted to assure specified quality.

Progress of Future Expectations

Bendix had not actually taken any steps toward changing the nonequity aspects of its Vartek agreement into a joint venture, but it was still considering this alternative.

Other than this option, which had to be exercised before January 24, 1980, no immediate projects were in the offing either in Yugoslavia or other countries of Eastern Europe. It was anticipated, at this stage, that the agreement would be extended on a year-by-year basis after the initial ten-year time period is reached in 1986. Bendix also hoped that the agreement would materialize into a larger project, although nothing concrete had been concluded.

NOTES

1. Lewis J. Lamm, article for Polish Foreign Trade Journal, 1977, p. 1.
2. Polish Machine Industry Offers 5(73) (1979): 8.
3. Ibid.
4. Eric W. Hayden, Technology Transfer to East Europe: U.S. Corporate Experience (New York: Praeger, 1976), pp. 48-49.
5. Lamm, p. 1.
6. Ibid., p. 2.
7. Hayden, Technology Transfer, p. 48.
8. Lamm, p. 1.
9. Hayden, Technology Transfer, p. 52.
10. Lamm, p. 3.
11. Ibid., pp. 9-10.
12. Jerzy Szumski and Janusz Piotrowski, "Bumar (Poland)—International Harvester (U.S.A.) Cooperation Is Now Over Five Years Old and Keeps Growing," Office of Polish Commercial Consul (Chicago, May 1978), p. 5.
13. Polish, p. 5.
14. Ibid.
15. Szumski and Piotrowski, "Bumar—International Harvester," p. 11.
16. Polish, p. 6.
17. Szumski and Piotrowski, "Bumar—International Harvester," p. 12.
18. Ibid., p. 18.
19. Polish, p. 5.
20. Ibid.
21. Szumski and Piotrowski, "Bumar—International Harvester," p. 19.
22. Ibid., p. 20.
23. Ibid., p. 15.
24. Ibid., p. 22.
25. Ibid., p. 17.

26. Business International S.A., "IH Reorders Priorities to Emphasize Asset Management Over Sales," April 15, 1977, p. 119.

27. Business, p. 115.

28. Szumski and Piotrowski, "Bumar—International Harvester," p. 13.

29. Ibid., p. 10.

30. Polish, p. 5.

31. Hayden, Technology Transfer, p. 54.

32. Ibid.

33. Polish, p. 4.

34. Ibid.

35. Ibid.

36. Business, p. 115.

37. Polish, p. 7.

38. Szumski and Piotrowski, "Bumar—International Harvester," p. 25.

39. Hayden, Technology Transfer, pp. 56-57.

40. Szumski and Piotrowski, "Bumar—International Harvester," p. 16.

41. Polish, p. 4.

42. Hayden, Technology Transfer, p. 57.

43. Szumski and Piotrowski, "Bumar—International Harvester," p. 22.

44. Ibid.

45. Polish, p. 8.

46. Hayden, Technology Transfer, p. 56.

47. Szumski and Piotrowski, "Bumar—International Harvester," pp. 23-24.

48. Ibid., pp. 24-25.

49. Ibid., p. 12.

50. Hayden, Technology Transfer, p. 57.

51. Ibid., p. 60.

52. Ibid., p. 63.

53. Donald A. Walczak, "A System for Transfer of Manufacturing Capability," unpublished manuscript (1978), p. 1.

54. Ibid., pp. 1-2.

55. Howard Wolff, "Polish Plant Ready to Roll," Electronics, October 14, 1967, p. 75.

56. Ibid.

57. Gus Prekopa, "Technology Transfer to Eastern Europe: General Tire's Experience in Romania," panel presentation at the 75th Annual Meeting of the National Machine Tool Builders Association, October 29, 1976, p. 6.

58. Ibid., p. 4.

59. Hayden, Technology Transfer, p. 96.

60. Ibid.
61. Ibid.
62. Ibid.
63. Lewis L. Lamm, address to the Foreign Trade Convention, November 18, 1975, unpublished paper, p. 7.

5

SUMMATION OF RESULTS, WITH CONCLUSIONS, LIMITATIONS, AND RECOMMENDATIONS FOR FURTHER STUDY

ASSESSMENT

The expectations and outcomes of selected corporate technology transfers to Eastern Europe have been presented in a case-by-case examination. However, an analysis needs to be made of these transfers in a collective sense, with the purpose of compiling suggestions for further nonequity approaches, especially in the Eastern European area. This analysis of results considers these transfers under two categories: descriptive factors and analytical factors. Descriptive factors, as laid out in Chapter 2, include:

Direct transfer of technology: product-process, skills/training tasks, language barriers, delays (specifically relating to the product or know-how), other technology-related factors
Indirect transfer of technology: construction, equipping and layout of plant or factory, raw materials acquisition and approval, other supportive capabilities
Environmental considerations: geographic factors, cultural interfaces, security issues, logistics, accommodations
Political arena: home governmental concerns, host governmental concerns, home labor union objections
Financial flows: payments, supplemental product purchases, incidental expenses
Leverage and control
Contract format
Expectations that did not materialize
Unexpected occurrences: rapport, change, product source, quality control, technological boundaries
Benefits to the corporation

Analytical factors, as identified in the eight technology transfers described in Chapters 3 and 4, include:

Technology: type, planning, staging
Training: needs, approaches, prior experience
Potential problems: disasters, delays, misinterpretations, change, development limits, competition, restructuring
Leverage: initial position, ongoing leverage
Contract preparation: negotiations, format
Facilitating aids: political involvement, labor unions, personal contact and support, realistic timing
Transfer rationale: financial compensation, other benefits

This chapter will conclude with an assessment of the limitations of these results and some recommendations for further study.

DESCRIPTIVE FACTORS

It should be pointed out that all the technology transfers considered were successful to both parties, at least in the short run. The technology was imparted to the receiving country, and the products or processes are, in every case, currently being produced (at least to some extent). Each country made the specified payments to the MNC as required by the stages of the transfer laid out in the contract. In terms of the MNC's bottom line, each transfer lived up to its promises, guarantees, and responsibilities. This refers both to the transfers that have already been officially terminated and to those that have yet to reach the concluding date as set by the respective agreements. Remaining MNC responsibilities are limited to updating provisions (this includes finished projects if extensions were granted on patents or other aspects), supplying additional models, obtaining a higher percentage of locally produced components, passing final tests, and/or providing additional approvals as required to meet certain specifications or trademark criteria.

Direct Transfer of Technology

Product-Process

The transfer of technology as exemplified by technical information such as carried by blueprints is apparently the least difficult aspect of the entire transfer process. Two aspects were emphasized by a majority of the transferrers: one, the more familiar the technology the more successfully it transferred; two, the more carefully planned the transfer the more easily it was accomplished with the least amount of disruption.

International Harvester, for example, found that the transfer of the more familiar crawler tractor technology was easier to transfer to Poland than its more recent wheel loader information. Honeywell found that certain product problems became apparent, especially during production. These difficulties were not limited to transfers to Poland but, rather, were first noticed on Honeywell's own production lines in the United States. Honeywell actually turned this particular problem area into an opportunity, however, when the additional attention paid to production in the United States resulted in the development of a systematic approach to teaching the method in Poland.

General Tire's radial tire transfer to Romania emphasized the disadvantages of transferring less familiar technology. GT's line of radials represented one of the company's newer technological areas. Consequently, GT had little knowledge of the performance of its tires over a long period of years, although the company certainly had adequate assurance of its radial tire's performance capabilities. However, some of the problems described in Chapter 4 might be traced to unfamiliar technology.

Those firms encountering "latest technology" problems all attributed them to the fact that the Eastern European countries insisted on the very latest version of the technology available. This was certainly not a criticism of this requirement. Despite any hardship or extra difficulty that might result, each firm had, for one reason or another, acceded to this request. In many instances, this concession was made by the MNC in order to obtain the contract.

It should be pointed out that once the agreement had been ratified, all firms were conscientious in keeping the Eastern European countries informed of the most up-to-date versions of the products. This benefited all parties concerned, particularly the U.S. firms (such as Clark, International Harvester, and Westinghouse) that anticipated using the Eastern European manufacturing facility as an additional source for its own international marketing needs. Honeywell noted that such strong emphasis on updating might cause some delays or slowdowns in the immediate production process, but in the long run both sides benefited from the latest advances or improvements in the product line.

Careful planning was stressed by a number of firms. The most detailed transfer was organized by Westinghouse (its computerized lists even included the number and placement of wastebaskets). The advantages of figuratively "freezing the process" were not confined to the transfer itself; Project Manager Walczak pointed out that Westinghouse was able to streamline its own production through such an intricate breakdown of tasks, equipment, and production steps.

Clark Equipment noted that careful planning was vital to attempting such a technology transfer, while Honeywell went one step further. It pointed out that not only must the transfer and training steps be carefully charted out, but primary insistence must be placed on absolute adherence to each and every instruction. Honeywell's insight had been gained secondhand through experiences of its Scottish subsidiary. Its resulting demands were apparently successful, since its transfer went smoothly despite the high complexity of its product. Westinghouse seconded this advice: careful planning was essential but had to be carried out to be effective.

The attitudes surrounding these technology transfers transcended monetary benefits. The possibility of repurchasing the products was one incentive behind the MNCs' concerns for high product quality, but not the only one. At least two firms supplied technology that could not be used in the United States (for example, Bendix in Yugoslavia). Others helped adapt the products to local conditions, as Westinghouse and International Harvester did in Poland. These firms also emphasized their desire to help Poland utilize the technology as much as possible; the adaptions were not usable back in the United States.

Technology flowed both ways as many of the firms became involved in joint venture projects that were expected to help the MNC to varying extents. Clark Equipment was one of the few firms to actually anticipate such a possibility in its contract. Several others, however, such as International Harvester, emphasized opportunities that might come to exist, such as making use of metric conversions as applied to product scaling in the United States.

In most cases, contracts contained clauses affirming the rights of the MNCs to any developments (including patents that might result from the technology transfer) and innovations created from their basic technology by their Eastern European partners. The purpose of such clauses might be interpreted foremost as an optimistic insurance policy that would protect the MNC's ownership rights should something of worth be derived. However, these clauses can also be viewed positively, as expectations that the knowledge generated by the partnerships would very likely lead to additional benefits.

Of all the joint ventures materializing from these technology transfers, International Harvester (IH) has apparently become involved in projects carrying the greatest magnitude. One project entails the metric scaling of certain parts of its vehicles, particularly the crawler tractors, which will, hopefully, be useful to IH in its U.S. manufacturing plants someday. IH was also working with the factory at Huta Stalowa Wola and Bumar in jointly developing a torque converter and a power-shift gear mechanism that would be valuable for both IH and the Poles.

Clark Equipment was involved in joint development projects in both Poland and Romania. Unfortunately, the Polish endeavor involving a metrically scaled axle became embroiled in a technical suit after Clark maintained that it should be paid a high percentage of royalties because the product developed by the Poles had been influenced by Clark technology. The Poles, on the other hand, took the position that their product was substantially different from Clark's, and that, therefore, only a very small royalty payment should be paid to Clark. Clark's joint development in Romania progressed with less controversy, and Clark, at least, felt that its Romanian venture exemplified teamwork. The Romanians designed the product, Clark agreed to help produce the prototype, and both partners were involved in the testing.

Neither Westinghouse nor Honeywell had become involved in a major joint development project, but this may have been due to the relative brevity of their contracts with Poland. General Tire, in Romania, was considering the development of a metric version of some equipment. This might be seen as a reverse technology flow, since information originating in Romania would be sent to the United States to help with the changeover to the metric standard—should such a change become necessary.

Product quality was crucial to each transfer. General Tire (GT) was the only firm among the transferrers considered here whose entire product, the radial tires, had yet to pass all acceptance tests at the time of this assessment, although some models or certain components were still to be approved on other projects. It should be pointed out, however, that the acceptance tests for the tires as laid out by Romania and GT did not simply involve a series of tests at one point in time but actually required durability runs covering more than a year. Furthermore, the tests in Romania had been severely delayed due to environmental factors.

The overall quality of the final product was a concern to all companies, but inclusion of the corporate trademark intensified this factor. IH distinguished between components that could be obtained from local or third-party manufacturers and those that had to come from IH or another approved source. In order for Bumar-built crawler tractors or wheel loaders to carry the IH logo, the vehicles had to incorporate the IH diesel engine, in addition to conforming to the above requirements.

Skills/Training Tasks

The virtually unanimous verdict from the U.S. firms was that the Eastern European trainees were far more skillful and had better technical expertise than had been anticipated. This was stated both by firms who had initial reservations regarding the available technological abilities of the trainees and by firms who assumed that the Eastern European workers possessed the necessary skills and background. Bendix appeared to be among the least concerned about training needs (despite the fact that the workers employed in the manufacture of friction materials for disc brakes had previously been employed as textile manufacturers—a highly divergent process). Ironically, Bendix found that a great deal of training was needed. However, Bendix was also highly complimentary about the ambition and determination the Vartek employees in Yugoslavia exhibited in mastering the new techniques.

In several instances, retraining or unlearning was a necessity. For example, IH found retraining necessary in Poland but essential in its Hungarian operation—in spite of the fact that the manufacturing operation in Poland was far more extensive. IH felt that the additional need for retraining in Hungary might be due, surprisingly, to the relative simplicity of the product. The complexity of the construction machinery transferred to Poland required Bumar workers to adhere strictly to IH instructions. In Hungary, the workers may have felt they already were familiar with manufacturing techniques required for producing agricultural machinery and, hence, did not follow IH recommendations as closely. It was also felt that IH did not adequately anticipate the extent of training that would be necessary in Hungary. This was a bit surprising, since IH had encountered a great need for certain training during the Polish project, and there was little reason to think the Hungarian project would be much different. In any case, IH's prior experience was not beneficial in this regard in that it did not provide additional guidance for other projects.

The GT project in Romania was another instance in which retraining was necessary to the transfer operation. In retrospect, General Tire representatives realized that the Romanians felt that they already knew how to produce tires because they had been manufacturing tires for a long period of time. The production of radial tires was a variation that did not seem different enough to warrant changing procedures.

Honeywell, as mentioned before, felt that part of its successful training experience in Poland was due to corporate insistence that all instructions be followed to the letter. The corporation also found that its training approach—instructing a small number of workers who then trained additional workers—was highly successful. The "each one teach one" approach tended to instill the skills as well as expanding the knowledge to others.

Prior contact with Western-based firms was apparently important in the quality control line. IH found, for example, that Bumar in Poland accepted the standard levels set by IH with more understanding than the Hungarians, presumably because Poland had become aware of Western quality control levels, while Hungary had not had much exposure prior to the IH project. To aid the quality control process in Hungary, IH reduced the number of inspection points from 27 to 15, to make the procedure as simplified as possible. Fifteen checkpoints appeared to be the lowest number of inspections that could be made without any resultant quality loss.

However, even prior contact with Western quality standards was not always sufficient. Overall, it took IH five years from the start of its endeavor in Poland until the first model was pronounced acceptable for the IH trademark. Clark found itself in a situation in which many of the initial workers on its projects in Poland were discharged because the production quality was not up to specifications. Unfortunately, Clark felt that the problem did not lie with the individual workers, but with the general adaption of Western quality standards. By the time new employees were trained, the entire group had, in reality, become more adjusted to Western quality demands. In general, Clark was eventually well pleased with Bumar reliability, product quality, and delivery commitments. Furthermore, it appeared that the training experience that Clark gained in Poland enabled the company to avoid some of the same problems in Romania.

Low morale and a lack of worker incentives were reputed to be commonplace throughout Eastern Europe, and many of the firms taking on these technology transfers had preconceptions of some of the conditions that would affect their training programs. However, except for some instances of low morale reported by Clark and IH, such factors did not seem to have a great impact on the training or skills transfer.

Some specific training problems were evident. Several firms observed that the high skills exhibited by the workers could be misleading because they did not necessarily indicate a general competency in all areas. For example, Westinghouse found it had to assess skills in each area and could not assume that particular expertise was always present. General Tire pointed out that the lack of positive incentives combined with the punishment accorded to workers making wrong decisions produced employees who refused to take on responsibility in Romania. Once all responsibility for decisions was removed, however, GT described the Romanians as hard-working individuals.

Training in certain processes was not the only type of technology or know-how that was transferred. Training in supervision and basic managerial principles was also found to be necessary. Westinghouse and IH both provided such training in an informal manner on their projects in Poland. Westinghouse found that it needed to instruct employees in cleanliness and safety standards. Pragmatically, it found that such standards were followed and enforced best when the reason for the regulations was plainly obvious. Clark, too, noted that environmental procedures, such as sealing concrete floors, were lacking in Romania and needed to be upgraded. IH even went beyond the level of supervisory training to provide marketing training to the Poles, and it helped RABA, in Hungary, sell some Hungarian products that were not even part of the IH project. Clark noted that attention paid to cost controls was finally being applied and that inventory control, extremely difficult at first, had improved as well.

The conclusion of a project was seen as an important phase of the entire operation. Westinghouse and its Polish partners created a film depicting the project's successful completion. As well, the project manager timed the departure of the Westinghouse employees so that the Polish workers were gradually required to operate the project on their own, staging it so that the transition was done smoothly and without great stress.

Language Barriers

Somewhat surprisingly—considering that all eight technology transfers involved at least two languages—no company reported any language difficulties over and above problems generally expected. Bendix felt that part of its success in avoiding misunderstandings in the language area resulted from the corporate practice of having a written record of all verbal interchanges or instructions. Westinghouse, with its computerized system, found that instructions written in both English and Polish placed at all processing stations were a great help. Westinghouse felt that problems often occur not from confusion resulting from language misunderstandings, but from not being able to obtain information easily—regardless of the language. To provide such information and to facilitate the production flow from area to area, Westinghouse developed a system of visible charts with production flows, expenses, and other information readily legible. This information was audio as well as visual, with a system of buzzers and lights worked out to provide warnings and signals more or less transcending any one language. In fact, Westinghouse seemed to feel that such a system of broadcasting basic information helped to reduce costs and streamline production as well as overcome language problems.

Delays

Delays cropped up on every project in one form or another. General Tire probably had the most severe delays, caused by the earthquake and the accident to the road-test vehicle. However, an earlier delay caused by the petroleum shortage in the United States just prior to the mid-1970s was actually the primary delaying factor. This petroleum shortage slowed the manufacture of equipment ordered by GT for the Romanian tire plant. As the equipment warranties were only for a specific period of time, GT found itself with the responsibility for repairing equipment whose warranties had expired. Subsequent disasters compounded the initial one, and time became a critical problem. International Harvester in Hungary, and Clark in Poland, also reported delays in obtaining orders of machinery for equipping the factories. Clark, in particular, encountered another instance of "cumulative" delays when the initial delays in axle production, caused by equipment shortages and training problems, were then affected by the impact of the recession. Thus, by the time production capability was finally on-stream, demands had diminished. Tooling delays also affected the IH project in Hungary.

Testing whether raw materials met requisite specifications was time-consuming for a number of companies. Sometimes just getting the raw materials caused even longer delays. To get its product on-stream, Bendix supplied some needed resins during the early years of the contract, assuming this would be a temporary situation and that such resins would be locally available shortly. General Tire in Romania, Bendix in Yugoslavia, and IH and Westinghouse in Poland all reported production delays resulting from the extended time required to test out the local raw materials. Westinghouse found that the raw material outages combined with the time needed to test the raw materials that were obtained doubled the delay factor.

Another factor creating some delays stemmed from the hard currency shortage in Eastern Europe. Honeywell, Westinghouse, and IH in Poland all reported delays in production caused in part because some components could not be purchased in quantity due to a hard currency shortage. Other delays were caused by special situations. For example, Westinghouse found that additional time was needed to achieve the level of cleanliness required for its intricate production.

Time projections were underrated in some instances. Clark found its initial time frame too ambitious in Poland, underestimating the time required to accomplish certain steps. General Tire, in Romania, found the same thing to be true. In both cases, however, the time frame had been set with five-year plans or other national objectives in mind—not just the project itself. In Poland, IH operated under the time schedule set by Bumar. The company suspected the time allotment was overly optimistic; IH was right. Clark's project in Romania was running a year behind schedule, but Clark did not feel that this was cause for concern. It felt that the original estimates were probably too brief because no production or other major problems appeared to exist in Romania. Westinghouse, too, stressed the importance of realistic time frames that allowed enough time to accomplish designated tasks.

Not all delays were caused by factors within Eastern Europe. Technology refinements back in the United States caused a delay both for IH and Honeywell in their respective transfers to Poland, as mentioned earlier.

On the other hand, Westinghouse felt that numerous delays were avoided because all equipment shipped to the project in Poland was collected, inventoried, tested, and packaged into containers in Pittsburgh. Furthermore, the equipment was held until the project in Poland reached the appropriate stage and was ready for its installation; then it was permitted to be shipped overseas. In the long run, this system carried a higher guarantee that the equipment arriving in Poland would be suitable and in working condition. Tighter control over the equipment inventory in this situation also helped to avoid losses, deter any major theft, and reduce breakage.

Indirect Transfer of Technology

Construction

Delays were encountered in the actual construction or renovation of the manufacturing facilities. This was most crucial for the U.S. firms supervising the construction, as in the case of GT, but it also affected the overall timing of the project in every instance. The firms that fared the best in these situations were those whose contracts had listed specific responsibilities for each party and had included stages to those responsibilities. Thus, Honeywell and Westinghouse, for instance, were able to encourage the Poles to complete their responsibilities in construction because the MNCs would not provide the next level of their technology until the specified construction had been completed. In such cases, the delays would cause Poland to incur additional expenses but would have little impact upon Honeywell's or Westinghouse's operations.

Equipping and Layout

The impact of delays on equipment warranties has already been mentioned, as has the advantage of pretesting and packaging equipment prior to shipment abroad. Several companies mentioned the advantages of purchasing new equipment because manufacturers' warranties usually accompanied such purchases. It was also noted that an accompanying advantage in such instances was that many equipment purchases could be negotiated to include machine setup and initial tooling in the purchase price. GT pointed out, though, that problems accompany new equipment (if the MNC undertakes the responsibility of providing this equipment), especially when the MNC functions as the middleman and uses outside suppliers. Since Eastern European agreements often have finite prices, any delays or increases in equipment prices beyond the MNC's control must be absorbed by the MNC. GT stresses the desirability, instead, of being the contracting agent, not the prime contractor.

One problem encountered by GT was a conflict between maintenance and repair duties. Equipment maintenance was the responsibility of the Romanians, but repair of the machines fell to GT. As Romanian maintenance was somewhat below the level usually desired by GT, repairs were more frequent. On the other hand, GT was gratified to note that the equipment seemed to be standing up better than anticipated under such circumstances. GT also encountered some instances of petty theft regarding equipment. The items removed were were, in most cases, not taken for the monetary value but for their unique qualities. Often these were objects that were not available for purchase in the local economy.

To avoid some delays in equipping the facility in Yugoslavia, Bendix not only provided a list of recommended equipment but also a list of substitutions, including items that might be available locally. As hard currency shortages were a universal problem throughout Eastern Europe, such a practice appeared to be helpful. In some instances, local sources might help avoid delays by obtaining certain equipment as well.

International Harvester encountered technology questions from the Poles relating to some of the equipment or materials bought from third-party suppliers. IH did not have the authority or information regarding such technology to transfer to the Poles. The Poles put what information was available to good use, however, as IH reported that Bumar itself constructed many of the tools used on the project. This created some delays at first, but it paid dividends in the long run, as expenses were curbed and repairs were facilitated.

Raw Materials Acquisition

Delays resulting from the additional time required to either obtain or test local raw materials have been described. The lesson appears to be that additional time should be allotted toward such purchases. Furthermore, initial testing alone apparently will be sufficient, as the quality level in Eastern Europe was described as fluctuating.

Most interfaces with raw materials pertain directly to the production involved in the technology transfer. However, GT encountered a situation in which it was to have first rights toward the purchase of certain Romanian raw materials in the petroleum area. This was initially laid out as part of the negotiating leverage directed toward helping the Romanians finance their radial tire technology purchase. However, when GT attempted to purchase these materials,

Romania refused and sold them to another bidder at a higher price than GT's offer.

Other Supportive Capabilities

Accommodations and logistics will be considered under environmental factors, but direct project support falls into the indirect transfer area. Westinghouse found that secretarial help was difficult to obtain, yet was crucially needed on the project. Dual language ability appeared most vital in this capacity.

Environmental Considerations

Geographic Factors

The geographic factor that most heavily affected the projects in Eastern Europe during this time span was the earthquake that hit Romania in March 1977. As this type of disaster was relatively uncommon in that region, the probability of such an occurrence was unlikely. As such, it was one of those times and incidents for which force majeure clauses are intended and, in GT's case, used.

Apart from this natural disaster, no other geographic factors affected any of the projects to any appreciable extent. The location of most of the plants or factories within larger industrial complexes actually acted as a kind of protection or insurance against many types of slowdowns or devastations, since heavy snows or other elements often would not affect production there as much as at a smaller or more remote facility.

Accidents are liable to occur anywhere, but the one incident that had a significant impact on a project involved GT's road-test vehicle, which ran off the road in Romania and was destroyed. Since this vehicle was road testing tires produced in Romania, another vehicle had to be prepared and the tests had to be started over—a factor that added to the delays already affecting the radial tire transfer.

Cultural Interfaces

In general, all companies seemed to appreciate the cultural heritages they encountered, as evidenced through their interfaces with the Eastern European countries. The people were described as friendly, and some of the Americans who had occasion to remain onsite in the region for an appreciable length of time reported fond memories of their stay.

Some elements, however, were not totally desirable. Working procedures were not as efficient as the MNCs would have liked, and incidents cut into normal working hours (such as extended sick leave, absences to attend funerals or celebrations, or even demands for flag-waving for officials enroute to or from the airport). On the other hand, Westinghouse found that the Poles did not take lunch breaks and, in this instance, the Americans had to adapt to having many snacks but no meal, per se, at noon.

General Tire found life in Romania more difficult than did most of the other firms involved in this study; however, it should be pointed out that GT's project required more Americans to live abroad (in Romania) for longer periods of time. Many of the other projects specified that visits by corporate trainers would last no longer than three weeks at one time. GT did distinguish between the official policy of the Romanians and the local citizens. On the local level, GT representatives found the Romanians to be admirable people. Four GT employees married Romanian women and brought their brides back to the United States. The approach of the Romanians was a problem, especially on the official level. GT felt that the official tone used by the Romanians was often threatening or accusatory. This conveyed a lack of hospitality as well as a reluctance by the Romanians to follow GT instructions.

Security Issues

Part of the inhospitable feeling in Romania was due to the high level of security in the country. GT representatives reported instances of harrassment, both on the roads and in their homes. The movements of the GT expatriates were carefully watched by the Romanians, and they even reported that they were afraid to make many friendly overtures to the Romanians for fear of being misunderstood or of getting their countrymen into difficulty.

Surprisingly, however, GT's experiences in Romania were not reflective of the impressions of other project representatives. Clark Equipment, also involved in Romania, had many supportive comments to make regarding the Romanians and their high skill levels. Clark did say that security appeared to be a problem in obtaining exit visas for the employees selected to come to the United States for training, but apparently this did not present an overly difficult problem. Poland and Hungary did not present any difficulties, and Yugoslavia appeared to be fairly open. Thus, it might be that the reason for the variation in impressions stemmed from the different projects in Romania. The rationale that an extended length of stay in a foreign country would produce a stronger negative attitude apparently was not always true, as other project representatives seemed to appreciate a country more the longer they were there. It should be pointed out that GT's project was faced with setbacks of a severity not affecting other projects. These problems may well have influenced attitudes on both sides.

In some instances, it did seem to come down to a matter of rapport. Initially, both Clark and IH interacted favorably with Polish representatives, but over the course of their respective projects, Clark's involvement with Poland tended to decrease while IH's increased. At the same time, however, Clark enjoyed a heightened rapport with Romania, while GT did not seem to develop the same relationship.

Logistics

Various logistical problems were frustrating but they were not actual project deterrents. Westinghouse noted that its own experiences might have been due, in part, to the particular contract arrangements that required all transportation between the United States and Poland to be on Polish airlines. As that airline's schedule was rather infrequent, frustrations mounted. Furthermore, Westinghouse was required by stipulations accompanying Eximbank regulations to utilize Polish carriers for a full half of all shipping on the project. This created some additional delays, as some of the Polish carrier schedules did not coincide with the Westinghouse shipping timetable.

Westinghouse, International Harvester, and General Tire all noted differences between the MNC and the Eastern European definition of <u>adequate</u>. Adequate transportation, under Eastern European standards, included available public transportation such as bus lines. However, the MNCs reported that such transportation was often not sufficient. Buses did not always run on schedule, often did not operate on weekends, and tended to not run often enough. General Tire resolved the difficulty by purchasing cars for its personnel in Romania; IH shipped over some International Scouts for its employees. Westinghouse rented automobiles in Poland.

Westinghouse also reported a problem with slow mail delivery, especially between the United States and Poland. To overcome this problem, a shuttle arrangement was organized. Employees traveling between the United States and Poland also carried official correspondence.

Accommodations

The same problem regarding varying definitions of adequate carried over into the realm of housing and living conditions. In retrospect, the MNCs felt that more specific descriptions of desired living arrangements within the original contracts would have avoided some of the disappointments. As it was, most firms tended to make some variations. Westinghouse had been assigned some apartments but these were felt to be unsuitable. Therefore, its employees shifted to a smaller, charming hotel, which provided restaurant and other services as well. Westinghouse employees found this arrangement ideal, as it was homelike, most serviceable, and permitted adaption to travel schedules. IH purchased some villas for its employees posted in Poland for the longer tours of duty and shipped in some food items that were unavailable in the local economy. It should also be noted that, for the most part, MNC employees were compensated handsomely for their work. Discontentment with certain other aspects was at least partially offset by their financial reimbursements.

Political Arena

Home Government Concerns

U.S. government approval actually exists on two levels in transfers of this type. The first level considers such technology transfers with regard to the political relationship between the United States and the other country. The second level involves technology that might be considered a security risk, often regardless of the country or region involved. General Tire included statements of U.S. government approval, a precaution with its roots in history, as Firestone had attempted such a transfer several decades earlier but had encountered both public and political antagonism to such an exchange.

On the second level, Honeywell was the only company among the transferrers involved in this study that was selling information that was potentially sensitive enough to require additional U.S. government scrutiny. Indeed, the initial contract was concluded subject to U.S. government approvals regarding the transfer of the sensor information. As it turned out, U.S. government permission was granted. Even with U.S. transferral permission, however, the sensor technology would not be provided until the conclusion of the five-year agreement. Honeywell worked out a system of controls whereby the technology was stored with a neutral third country as a precaution during the five-year period. This was a security measure against the intervention of political influences that might prevent the eventual transfer, despite initial U.S. government approval. The five-year time limit was completed without any intervening factors, however, so Honeywell was able to transfer the up-to-date sensor technology directly from its U.S. headquarters. Such a storage arrangement appears to be a highly feasible and a viable method for future technology transfers of a similar nature.

The spread of potentially sensitive technology from Eastern European countries to the Soviet Union is an issue open to interpretation. Most of the technology transfer agreements contained stipulations that the technology was not to be transferred to additional parties for reasons that ranged from political aspects to those of patent and trademark protection. However, reinforcement of these stipulations, at best, was only for the time period indicated on the contracts. Furthermore, control over the use of this information appeared limited, as enforcement would be limited to international legal measures of uncertain strength. Several companies did admit to concern over wrongful use of their technology, both in political and economic senses, but apparently have taken the gamble that their Eastern European trading partners would abide by contract regulations.

Only one company, Westinghouse, reported an undue amount of U.S. government red tape involved in obtaining an export license due, it would seem, to the fact that this transfer was somewhat unique in the type of information being transferred rather than sensitive in a strictly political sense.

Host Government Concerns

The cooperation received from the host governments in arranging and implementing these contracts was far more supportive than originally anticipated by the majority of the MNCs.

This support was initially most evident in situations in which the Eastern European country had sought out the U.S. company. In such instances, even the negotiating time was often substantially reduced, as in the case of Clark Equipment and Poland, or General Tire and Romania. Bendix was reportedly amazed that the approvals in Yugoslavia apparently did not even require permission at a high government level, and that the contract was able to be concluded at the factory level. In this instance, once assured that Vartek would honor the contract with all its accompanying regulations, Bendix was ready to proceed with the transfer. Westinghouse and IH both described Polish officials as extremely cooperative throughout the operation of the transfers.

General Tire found that government participation at higher levels helped to keep its radial tire technology transfer more efficient, since officials with necessary authority were willing to step in if the occasion warranted.

Westinghouse encountered a unique situation in that its subsidiary in England registered a protest against the technology transfer to Poland—not on the basis of the technology itself, but because of some of the equipment being sold to Poland for the manufacturing operation. Their protest stemmed from the type of blast furnaces suggested in the equipment list, because the English subsidiary felt that its product line should have been suggested.

Home Labor Union Objections

It was expected that the major union objection to the technology transfers would center around the loss of potential jobs for Americans. Because of this, several of the firms involved in these nonequity transfers pointedly noted two aspects of the transfers that actually benefited U.S. workers. The first item was that in such transfers, the question of creating a facility in the United States or outside its boundaries was not an issue. In each instance studied, the firm did not have (or had not apportioned) the capital needed to construct such a facility—hence, the likelihood of building the plant in the United States was not even a possibility. Second, the sale of U.S.-made components and, in some cases, whole or subassemblies, was substantially increased, thus increasing U.S. exports and benefiting U.S. workers.

Several firms, such as Westinghouse, had expressed these advantages to the unions with the view of preventing dissension, and no firm reported overt union problems over these technology transfers.

U.S. workers did cause one problem, however, though it was not promoted officially by any labor union to potentially jeopardize the project. International Harvester found that some of its individual employees had strong objections to dealing with Communist countries and were attempting to sabotage the operation by putting anti-Communist literature into boxes of components being shipped to the project in Poland. The Poles, unpacking these components, were upset and angered by these messages, thinking that IH did not really want to engage in the technology transfer. It was not until IH discovered what was happening and put a stop to the propaganda campaign that relationships became better.

Financial Flows

Summarizing the outcomes of the financial flows from the projects resulted in several observations, most based on simple common sense. For example, accurate timing was crucial in achieving projected profit margins. Exceptions to this statement, however, were the delays in production that actually aided the financial flows for several companies because additional products and components were purchased to compensate for the losses in production.

Payments

All arrangements between the MNCs and the Eastern European countries had payment schedules that were divided into at least two major area: initial disclosure fee payments and ongoing technical assistance fees or royalty payments. In most cases, the initial fees constituted the major bulk of the entire financial payment, providing a less risky situation for those firms receiving the payments. In all instances to date, these payments had been made as stipulated. Westinghouse did encounter one problem because the U.S. banks that would be handling the transactions had not been specifically indicated—with the result that Westinghouse, at first, literally had to chase down its payments from Poland. This may have been partially the fault of the contract terminology, but more likely it was the result of a new project where operations had not yet been standardized, and of the late arrivals of approvals both from Warsaw and from the United States. Next time, Westinghouse noted, it will stipulate the banks to be involved.

General Tire's operation in Romania was somewhat unique from other projects in two respects. First, a larger proportion of its payments were due during phase two of its operation than in most other projects; and second, GT had been beset by expenses caused by the earthquake and other events that delayed completion of the acceptance tests. Subsequently, the shift into the second part of the project had not yet occurred. Due to the unexpected disasters, GT was disappointed in its financial flows from the first phase of the project but was hopeful that the second phase would be better as soon as the acceptance tests were completed.

Several firms emphasized the desirability of an escalator clause in the contracts regarding purchase payments for both equipment and materials, as well as the payments themselves. This escalation might be tied to current price levels, or be an annual increase percentage, but its mere existence would provide the MNCs with an opportunity to guard against changing conditions. The difficulty with the inclusion of such a variable clause is that many contracts with countries in the socialist bloc must have finite financial limits laid out for the entire length of the contract. Such a situation, however, often places the MNC at a disadvantage since prices usually rise. However, in one instance, the market price actually dropped below the price specified in the contract, but both parties were required to follow the prices as stated in the agreement.

Another clause that is sometimes included in Eastern European contracts is the holdback clause. This provides the host with the option of not paying a final part of the fee until all responsibilities have been met by its contractual partner. Since something is usually subject to criticism in projects of any great magnitude, the holdback clause often becomes an obligatory extra fine or payment. Companies noted that if a holdback clause were required, it was best to have a very specific list of the responsibilities and duties of both partners. It is particularly important that one party not be jeopardized because of the actions of the other party. GT felt that its arrangements with Romania tended to balance out as, on the one hand, its contract did contain a penalty clause but, on the other, GT avoided a holdback stipulation.

The financial flows generally reflected the balance of controls or leverage of the projects. Westinghouse and Honeywell each had contracts that clearly laid out all responsibilities for the Poles as well as for the domestic companies. Furthermore, in their contracts all expenses were the responsibility of the host country (this was also the case for Bendix in Yugoslavia). At least a minimum royalty-type payment was required for each year of the project, whether production was in progress or not. Thus, these companies were able to control their parts of the operations, and keep them in balance with a minimum of expense. International Harvester was to receive a royalty-type payment from both Poland and Hungary, but not until production was under way in each case. As delays were encountered, requiring additional IH employee time and travel, these expenses cut into IH's financial estimates for the projects.

Very few cost-cutting hints were provided by any project managers, except for the suggestions by Westinghouse to make the expenses as visible as possible. Westinghouse also felt that tight inventory control combined with pretesting of all equipment helped to keep costs much lower than would otherwise have been the case. Even with its careful control, however, Westinghouse found that it was difficult to isolate travel expenses from living expenses, and adopted a per diem cost approach instead.

Clark Equipment, in Poland, found that inventory problems affected its financial basis. Clark had the responsibility for selecting the axle models to be produced. Clark found that a relatively small number of numerous models were actually in demand, although its assembly line techniques encouraged greater production of fewer models. Clark's financial flows in Poland were affected by more than just the number and types of axles being produced, however. The timing of its Polish operation, with production slated for the mid-1970s, coincided with the recession that hit Europe about that time. This slowed both construction and demand for construction vehicles. Since the Clark axle was a major component in these vehicles, the diminished demand had a direct impact on the profitability of the entire project. This was not always true, though, as its project in Romania was just far enough behind the Polish one to escape the heaviest impact from the recession. Demand within Romania was such that Clark products were purchased even prior to the start-up of production. Ironically, the Romanian project was bringing greater financial flows to Clark than the Polish project, which was of a much greater magnitude: a reminder that results, even in well-planned projects, are not always predictable.

No company expressed a major concern with receipt of the royalty or technical updating fees that would be paid during the latter portion of the agreement. General Tire was perhaps in the riskiest position, as its technology would have to be completely transferred before production could commence. Thus, the technical service contract payments could be seen as updating fees, but GT's leverage at this point would not be very high. Many firms had opted for specific payments instead of actual royalty fees, which are usually computed from product sales or production quotas. Bendix, one of the few firms to receive actual royalty payments, expressed no concerns regarding the auditing or bookkeeping involved. Since the brake materials were to be installed as original equipment on cars being produced in Yugoslavia, Bendix would have an indirect method of supplementing its other production records by determining the number of cars produced. In comparison, GT admitted that it had no knowledge of the actual number of tires being produced in Romania.

Supplemental Product Purchases

Product purchases from the domestic company varied from project to project. Poland's purchases of Honeywell subassemblies were far greater than ever anticipated, and Poland was glad to be able to purchase the units, which acted as training aids while also supplementing production. Furthermore, Poland requested that the ceiling be raised on the number of sensor units it was allowed to purchase, despite the fact that the country would be provided with the sensor technology. Tooling machines to manufacture the sensor units were anticipated to be very costly, and continuing to purchase the items from Honeywell was seen as a less expensive alternative. Honeywell was very pleased with its overall sales. Other firms such as International Harvester in Hungary reported that certain items—for example, component sales—were higher than anticipated, compensating for other sales that were lower than originally projected, particularly those of complete units or total vehicles.

Westinghouse reported that its project with Poland benefited some of its third-party suppliers. By bringing those vendors into Poland to set up the equipment, their particular product lines were introduced in Eastern Europe. Some of the vendors were subsequently invited over again for trade shows and other business, totally apart from Westinghouse.

Incidental Expenses

Generally, purchases of transportation vehicles and accommodation supplements exceeded corporate expectations in this category. However, this applied, on the whole, only to firms involved with on-site responsibilities over a long period, not to those whose involvement was concentrated into shorter time periods. Westinghouse and IH found that more trips than expected were necessary between the United States and Poland for their respective projects, requiring extensions of both time and expense estimates.

Leverage and Control

The firms involved in this study did not believe that equity arrangements provided any essential elements of control in the operation of a project over and above those found in nonequity arrangements. In fact, the two firms with options to shift into joint ventures, General Tire in Romania and Bendix in Yugoslavia, opted not to take that route.

Selection of the nonequity arrangement was probably made more from necessity than preference in some instances. Several firms noted that they had been disinclined to finance a project of the magnitude of the ones in Eastern Europe; but some firms, such as Clark Equipment, stressed the desirability of such nonequity arrangements. Clark's opinion was that equity involvements (referring to joint ventures) did not really provide the nonhost partner with an adequate control basis if facilities were located on foreign soil. This was felt to be true even if the arrangement was one including majority interest. Instead, Clark felt that leverage ought to be built into projects on other bases, regardless of the financial arrangement. Clark considered the gradual transfer of technology to be one method for gaining control—holding the crucial formula until the very end is an example of this type of leverage.

Both Clark and General Tire assumed that the inclusion of their corporate logos and trademarks in the technology package would provide additional leverage; but both found this to be exaggerated in importance. Clark's trademark was revered throughout Poland and Romania, but since its product was actually only a component of an entire vehicle, the visibility of the trademark was, for the most part, lost to the general public. In Romania, General Tire had set up the agreement so that the acceptance tests had to be successfully completed before permission to use the GT logo would be granted—thinking that this would be an additional incentive to complete the acceptance tests. Instead, Romania was apparently more inclined to manufacture radials without a trademark—and without any royalty payments—than it was to pass the tests and use the GT logo.

Another form of leverage was evidenced by an arrangement in which the MNC agreed to undertake marketing or other selling responsibilities. Clark noted that in another agreement (not studied here) continuing access to Clark Equipment Company's distribution system constituted a highly desirable entry that gave the company leverage throughout the course of that particular agreement. The same type of control was found in this study, particularly between Clark and its Romanian contacts. The marketing and sales aid that Clark was able to provide to Romania gave the corporation leverage and also encouraged greater teamwork, since increased sales financially benefited both parties. This area of leverage is not limited to marketing expertise but applies to any benefit that the MNC might be able to offer the interested party. GT holds the technology for tires for off-the-road vehicles, for which Romania has expressed a great interest. Although GT is desirous of selling this technology to Romania, it has taken the position that such a transfer will not be considered until the current project has shifted into phase two, with all acceptance tests completed.

One of the more successful incentives for keeping projects within designated time frames had to do with structuring the contract so that the U.S. firm could go into a waiting pattern until the host partner had completed certain responsibilities, such as construction or equipping. This arrangement, however, had to be constructed such that accompanying expenses were also the responsibility of the host country or company—not the MNC. Honeywell and Westinghouse provided good examples of this in their arrangements with Poland, and Bendix used a similar format in Yugoslavia. Bendix actually went one step further in requiring annual royalty payments each year, even before production was operational. These payments provided an additional incentive to have the manufacturing operation under way as soon as possible to help offset the outflow of capital. This arrangement also contained incentives for Bendix, because the sooner production was on-stream the sooner its payments would exceed the minimal level.

The buy-back or counterpurchase might also be considered a form of leverage, particularly in the area of quality controls. However, this position also includes the possibility of misunderstandings, as IH found on its Hungarian project. IH was rejecting the finished agricultural products on the basis of inferior quality. Hungary, on the other hand, misinterpreted these actions as an indication that IH did not want to keep its portion of the agreement by repurchasing the goods. Refusal on the basis of inferior quality was one justification that IH was permitted to use. The misunderstanding apparently stemmed from the fact that Hungary was not very familiar with Western quality standards.

Another form of leverage is found in the MNC's suggestions or recommendations for plant equipment and layout. If the firm is permitted to make suggestions without having to undertake additional responsibilities such as overseeing actual construction, equipping, or even tooling, this apparently is even better. General Tire's comments that the company would prefer not to be the prime contractor in future agreements bears this out. The power to suggest or recommend equipment choices has two advantages. On one hand, the company is able to suggest equipment that it would most prefer to see installed in the facility. On the other, working with new equipment at the start of a project provides a buffer time when the machines are under warranty but the MNC is relieved of repair expenses. This gives the company a great advantage toward achieving high quality production from the start, and often it even includes equipment representatives responsible for machine setup and some tooling. Thus, in effect, this provides additional help without more expenses.

One of the best leverage positions to be in is that of sole proprietary owner of the desired technology. Honeywell was in such a position, especially with regard to its sensor technology. However, some firms are apparently able to achieve many of the results that might be accorded to monopolistic situations through strategic bargaining. General Tire, for example, was hardly sole proprietor of radial tire technology, but other U.S. firms were less inclined to become involved in a nonequity technology transfer. European firms were not as attractive to Romania, perhaps for political reasons relating more to the attainment of most-favored-nation status than to other, technical reasons. At the same time, GT avoided a required counterpurchase aspect to its agreement because the European firms also did not want to accept such an arrangement. Thus, GT was able to receive benefits and bargaining advantages not always available to firms not having sole rights to certain technology. Therefore, this was a result of several factors, mainly political justifications that transcended the firm itself, and inside knowledge of competitor inclinations. GT has emerged from its current technology transfer with valuable experience and advice. This might be summarized as "document everything," supplemented by the admonition, "always be prepared to stop the transfer with a minimum of loss."

Contract Format

Ellison's observation that the more general type of contract was the most successful approach to Eastern European agreements was not borne out by the experiences of the firms in this study. In fact, Westinghouse, whose entire transfer was extremely detailed, apparently had the fewest problems. General Tire had a fairly large agreement (that contract covered more than 300 pages), but considering the magnitude of the transfer and some of its outcomes, even more detail regarding certain aspects of the transfer would have been desirable.

Perhaps it would be more correct, however, to say that certain leverage aspects were missing from GT's arrangement, particularly with regard to tire production prior to completion of the acceptance tests. International Harvester, Honeywell, Clark, Westinghouse, and GT all stressed the opinion that the more specific the contract, the better.

Logistics and accommodations were one area where all firms felt that expectations should be extremely detailed, as terms such as adequate varied between partners. Specific details included such things as room size, bathroom facilities, and transportation schedules.

Westinghouse also listed those items that the company was not responsible for—a list that could conceivably become too extensive to be practical if not done carefully. In this instance, however, it helped to clarify tasks and generally aided the functioning of the project.

This procedure would have been especially beneficial for GT, because the Romanians took the position that if GT had not stipulated what it would not do, GT had to do it. General Tire's feeling was that its real discussions with Romania did not commence until after the contract had been signed. Many of these discussions revolved around interpretations of the contract's wording, with literal meanings being taken seriously regardless of the accompanying context. GT's warning regarding the earlier negotiations was that throughout the bargaining sessions, extreme caution had to be taken to ensure that no words, figures, or amounts were changed, proportionwise, unless those changes were desired. On the other hand, GT felt negotiators on both sides had to be willing to compromise or to exhibit a willingness for give and take. Otherwise, GT notes, negotiations tend to stalemate.

The U.S. firms continued to emphasize the fact that while a corporate name or a certain reputation might be the initial reason for gaining entry to a certain area or country, personal acquaintanceship was vital once negotiations were underway. Contracts were viewed as agreements between friends, and that friendship had to be forged before the contract was ready to be signed.

Negotiations played a role over and above that of drawing up the final contract. Several firms noted that the rapport generated during long negotiating sessions often paid off in more efficient functioning of the agreement. As an example, Clark Equipment's initial negotiating session with Poland was brief (brief in comparison to most negotiations with countries in Eastern Europe), but proper functioning of the contract then required more personal interaction. This differed from Clark's agreement with Romania in which the additional time spent laying out contract details provided a stronger operational basis once the project was instigated. Overall, however, Clark still feels that one of the benefits of doing business with Eastern Europe is that once the agreement is finalized, it is followed.

Expectations That Did Not Materialize

Most of the firms involved in this study had some preconceptions of the problems they would probably encounter in a nonequity technology transfer. Some of these expectations did not occur at all, while others had far less impact than originally anticipated. Honeywell had foreseen language problems, since the transfer involved countries with different languages, but no undue language problems occurred in Honeywell's transfer. A number of firms had been primed to expect low worker morale and accompanying quality control problems, but these aspects did not cause much difficulty. Instead, the companies were universally amazed at the technical skill and expertise possessed by the employees of their Eastern European partners. Complaints about the trainees mostly appeared (if at all) at the start of the project, as with Clark or IH in Poland. Earlier reports such as Hayden's found corporate officers somewhat discouraged at times. However, over a longer period of time, these training problems were apparently overcome. In retrospect, these aspects did not appear very consequential.

Clark had anticipated more trade and interchanges with other countries in Eastern Europe but found that this was not the case. Essentially, this trade gap was due to the hard currency shortage that faced many Eastern European countries. To overcome this deficit, trade was sought with outside nations or regions. For Clark, this turned out to be a disadvantage, since part of its corporate projection for the future was that the axles produced in Poland and the transmissions produced in Romania might, someday, be combined with other components produced by other nations within the Soviet bloc. This remains a future possibility, but it is not considered to be as viable as once projected.

Whereas the technical abilities of the workers in Eastern European projects were more skillful than generally assumed, several firms reported that their initial expectations as to worker familiarity with specific manufacturing techniques were not fulfilled. IH, for one, had anticipated that the workers in Poland would possess quality control knowledge and perseverance, both of which were initially lacking. In several instances, the workers themselves apparently believed that the technology would be familiar. This misconception was apparently the basis for some initial sources of conflict when the U.S. firms felt that the Eastern European workers were not following instructions. However, when the U.S. firms carried this sense of independence one step further and required that decisions be made by the workers, no decisions were made, as GT found out in the Romania project.

Future expectations, in general, materialized far more slowly than originally expected. Firms that had completed their specific projects, such as Westinghouse and Honeywell, reported that future projects had not been ratified, although they intimated that some negotiations were in progress. Clark described the situation as "more contacts, but fewer agreements." The firms with projects still in progress also spoke of future possibilities, but in a less comprehensive sense than had been the case in the early 1970s. There seemed to be several causes for this reevaluation. First, some projects had been discussed but not selected. One example was Clark's hope of signing a construction vehicle contract in Poland, which was awarded, instead, to IH. Another example was a contract that Westinghouse hoped to receive; it went to someone else. Second, the recession and economic slowdown throughout Europe in the mid-1970s scaled down many aspirations and reduced demand in several industrial areas. The third reason related to a more realistic appraisal of the market potential in Eastern Europe, with some firms scaling down their original estimates.

Companies that have refused to sell their technology reason that such a transaction literally creates a competitor. Some of the companies involved with this study acknowledged that such a possibility did exist whenever technology was transferred, particularly when the firm's most recent technology was involved. However, none of the companies in this study felt that their agreements would turn out to be competitively threatening. Those companies with counterpurchase or buy-back options hoped to use their international contacts to help meet their marketing needs. Others felt that their research and development would keep the domestic companies ahead of their Eastern European counterparts. Several firms had designated regions or areas where their Eastern European partner would have full marketing and distribution rights to the product, thereby indirectly noting other distributors or areas where the MNC would retain marketing prerogatives.

Surprisingly, the one discordant competitive note that sounded during this study came from Bendix, but not Bendix, U.S.A. One of the Bendix European subsidiaries protested that the Yugoslav project would produce a direct competitor for the subsidiary. Bendix U.S.A. did not feel that was likely, particularly since the technology that was transferred had not included any marketing commitments. The European Bendix subsidiary had an experienced sales force as well as access to Bendix marketing information and support—factors that should definitely place it far ahead of the Yugoslav operation.

Anticipation of political problems was not overtly expressed, but several companies did describe some initial uneasiness regarding their involvements with the communist system. No firms reported any problems in this realm, however, with the exception of GT's examples of minor harassment. In fact, most of the firms reported that while they assumed that some of the trainees who were selected to be sent to the United States for training were members of the Communist party, no mention was made of it in any training situation.

Unexpected Occurrences

Most of the unusual, unexpected events, such as the earthquake in Romania, have been covered under other topics. However, it appears valuable to list some additional corporate observations here.

Rapport

The desire for a corporation's particular technology was the general incentive behind each of the initial agreements, but future possibilities stemming from those agreements hinged upon the rapport that developed from those technological offerings. Interestingly, Clark Equipment had developed a supportive relationship with Romania, while IH became involved with Poland. Clark encountered few of the problems that plagued GT in Romania, which tends to emphasize personal factors over cultural or even technological ones. Clark also noted that corporate commitment to such technology transfers was also essential in establishing a high degree of rapport among members of the MNC's team and that this commitment made involvement with others easier and smoother. Westinghouse noted that, regardless of future endeavors, rapport was vital in accomplishing the current project; teamwork was viewed as essential.

Change

In retrospect, Clark Equipment noted that technology transfers often require the recipients to adapt to another group's working habits as well as to become familiar with a different set of machinery or equipment. When these changes are relatively large, such as in the technology Clark transferred to Poland, all groups require additional time and effort just to adjust to the changes. Clark felt that processes such as adaption to change (caused by extensive overcapitalization involved with massive equipment purchases) should be built into the technology transfer plans. This information actually might fall under the know-how category, but it transcends the process information.

Product Source

Emphasis was placed on buy-backs or counterpurchases, especially in areas where hard currency was in short supply. However, such an arrangement often is less convenient for the firm transferring the technology because it means that payment (or partial payment) is postponed until production is under way. Thus, most companies prefer outright payments, at least for the disclosure fee. In these studies, IH in Poland had only an optional counterpurchase arrangement, but it was beginning to see the long-run potential of having an additional product source without having to finance or build a facility. The growing consideration of this sourcing possibility speaks well of the project and certainly provides an additional reflection of the skill of the Polish workers.

Quality Control

Maintenance of standards was one of the primary focuses of most of the transfers studied here—regardless of the inclusion of the firm's trademark in the agreement. It was assumed that products from the company's other divisions, especially those from the United States, were of uniform and consistent quality. However, Clark Equipment found that its own quality control standards had some problems. When it supplied some items to its Polish project for use in the training of quality control standards, its own products did not measure up. As a result of this discovery, Clark took steps to improve its own quality control checkpoints.

Technological Boundaries

Several firms reported requests by their Eastern European partners for technology that was either necessary or desirable but out of the MNC's personal domain. In such instances, the technology simply could not be provided.

Several contracts had provisions for a two-way use of technology developed from the initial project. This helped to assure the MNC that that it would have rights to the research and development taking place in Eastern Europe that stemmed from its technology. It also provided a basis for royalty payments back to the MNC, should use be made of the domestic company's technology over and above the uses specifically designated in the contract. Clark, however, had a conflict with its Polish partner over the limits of this technology. The Polish engineers maintained that their adaption had actually utilized very little direct technology from Clark Equipment. Clark, on the other hand, felt that its technology had provided the inspiration behind the new developments. In this instance, the controversy waged over the amount of royalties that Bumar owed to Clark. The solution was a compromise between the sums requested by each party. If the compromise had failed, Clark was willing to resort to arbitration—which would have been the only instance to date where arbitration had been used in any of these particular projects. This incident contrasted with Clark's involvement in Romania. In the latter case, Clark's joint development was written up to provide Clark with rights and financial compensation as the jointly developed transmission was sold.

Corporate developments brought about some changes during the course of the agreements. IH totally revised its international organizational structure, resulting in a different group having charge of the Bumar project in Poland. At the same time, Poland shifted some of its economic structure. However, neither change appeared to cause any problem and, if anything, the shifts appeared to bring relationships between Poland and IH closer than before.

Recall that the original opening for the Bendix technology transfer to Yugoslavia occurred through Bendix's French subsidiary, DBA, which was negotiating a disc brake project with Yugoslavia. When DBA realized it did not possess the technology for producing the friction materials for disc brakes, the U.S. corporation had become involved—such opportunities are one of the advantages of the MNC, which has a wider range of technology and resources at its disposal. However, these wide-ranging contacts can also create a point of contention. Bendix acquired another subsidiary after the contract with Yugoslavia was in effect. This subsidiary, located in Europe, possessed the technology for a different model of friction material. Yugoslavia requested that this technology be provided, since the contract with Bendix stressed the point that all updated and advanced information relating to the technology was to be transmitted to Yugoslavia during the course of the agreement. Bendix, however, felt that such a provision certainly applied to its immediate operation, but not to its separate subsidiaries. In this instance, a compromise was reached as a settlement. Bendix provided the information, but training in the production of this other formula was done at a European location rather than in the United States. This training would provide the necessary information relating to the additional technology, but it would not provide any supplementary information regarding additional processes or other information.

Benefits to the Corporation

Bendix, Westinghouse, Honeywell, and IH (especially in connection with its Polish agreement) all stressed financial gain as their primary reason for entering into nonequity agreements with an Eastern European partner. The outcomes of those arrangements (at least thus far) have brought their expectations to fruition. Clark Equipment in Poland, and General Tire in Romania, expressed disappointment over the slower rate of their monetary rewards or product source. On the other hand, Clark was well pleased with the financial flows resulting from its involvement in Romania.

Alternative product source capability was a benefit gained by several firms. Clark and IH, in particular, stated that this possibility was among the primary benefits of their transfers.

Obtaining a toehold in Eastern Europe was cited by most of the firms as an accompanying reason for the transfer, but only Clark and IH actually concluded a second agreement with another Eastern European country. Bendix ran into an impasse regarding its Soviet contract, and other negotiations had not culminated in tangible projects. Thus, while the firms all felt they had expanded their corporation's reputation in that region of the world, increased involvement would apparently take more time to develop than originally expected.

Additional advantages materialized for some of the companies. Honeywell sold more components and subassemblies than ever anticipated; IH began to realize product source capabilities through its Polish and Hungarian projects; and Clark was finding its joint development projects, especially in Romania, to be both financially and technologically beneficial. Westinghouse's computerized approach to the transfer helped streamline its operations back in the United States. Clark found that the extraneous information gained from its Eastern European partners was beneficial, and several firms were benefiting from the joint research and development that was taking place.

ANALYTICAL FACTORS

Certain aspects stand out when studying nonequity technological transfer projects with Eastern Europe. Some of these factors are particularly helpful in the progress of the technology transfer; some are detrimental. In some instances, certain factors appear crucial, in that their absence might render a project undesirable or unprofitable.

Technology

Type

Certain technological characteristics appear to affect the ease with which a transfer may take place. The transfer of the information is perhaps the easiest step of the entire transfer. As such, those firms able to precisely define their technology would be the most likely to successfully transfer that information to others. Westinghouse illustrated this feature through its elaborate computer printout transfer method. The fact that the Westinghouse technology was extremely complex and intricate makes this successful transfer all the more commendable. It also emphasizes the concept that the procedure used in transferring the information is more of a factor in determining the success of the transfer than the complexity of the material.

Familiarity with the technology is also highly important for the MNC. International Harvester, Honeywell, and General Tire had each promised the latest developments in their respective technologies, but each encountered difficulties partly because the technology was new and unfamiliar even to them. This is not to say that corporations should not attempt to transfer their most recent technology. The long-run advantages are significant: the corporation helps to ensure component interchangeability, and it maintains a more viable repurchase or product source option. However, additional care should be taken in such instances to gain as much understanding of the information as rapidly as possible.

Planning

Though it sounds redundant to say that the best planned transfers appeared to be the most successful, the MNCs virtually universally emphasized this point. In this context, planning ranges beyond the point made above of having a thorough knowledge of the information to be transferred to include a comprehensive, step-by-step approach to the transfer. This planning should incorporate contingency actions as well as detail all phases and aspects of the transfer. Again, Westinghouse, with its emphasis on detail, probably used the most intricate approach through its computerized checklists, but other firms supported the fact that the more planning done in advance, the more likely the transfer was to go smoothly.

Another issue here, which relates both to the type of technology and the planning involved, concerns the amount of information to be transferred. Complexity does not seem to be as crucial as the actual extent of the operation. Both Honeywell and Westinghouse transferred highly intricate information, but each transfer was relatively contained. International Harvester and General Tire were responsible for much larger operations and found that the magnitude of such transfers carried a proportionately larger margin for potential problems. In such instances, planning would also be required to be far more extensive.

Staging

Several aspects relating to different stages of the transfer were highly important. Wallender's observation that different approaches should be considered and employed by the MNC at different points during the transfer was borne out by several firms. Essentially, the approaches consisted of early contacts to establish rapport and facilitate acquaintance-making sessions, followed by a period in which the information is disseminated with close control over its initial applications. After this point, however, when it has been determined that the essential ideas have been communicated, firms recommended a gradual weaning away, placing more and more responsibility in the hands of the local employees. Westinghouse made a film of the completed transfer showing the products being successfully produced in Poland as a tangible illustration of the project's conclusion. In several instances, the conclusion of the project was softened in that, while the project itself was over, ongoing involvement in the form of patent or trademark extensions or technology updates were in progress, so the involvement was not abruptly cut off.

The length of time needed for the transfer was a factor, but it did not appear to be a clear-cut advantage or disadvantage. The shorter projects were certainly more manageable, but in such instances additional agreements or other projects had not been forthcoming. The longer projects were at greater risk with regard to payments, and they seemed to be accompanied by more difficulties at first (but this would be largely attributable to the larger project size). Furthermore, the longer projects appeared to generate greater rapport and more likelihood of additional involvement.

Thus, it would appear that some general rules of thumb should apply in most technology transfers to Eastern Europe, including the following points:

The smaller, the better
The shorter, the better

As is often the case, however, these examples of technology transfers also support the opposite position. IH's technology transfer to Poland involved entire vehicles, a huge operation, including renovation of facilities and enormous equipment installation—and IH is experiencing one of the most positive relationships with Poland. Its training difficulties were basically limited to the first few years. General Tire's ten-year project (plus building time) with Romania was one of the lengthier ones; but, in retrospect, GT's relationship with Romania improved over that time period. The unusual difficulties encountered might have jeopardized the project completely if the time frame had been much shorter (on the other hand, the shorter the time frame, the less probability that such difficulties would have been encountered in the first place). As it was, the teamwork required to succeed in the face of adversity was building greater rapport between GT and Romania. Therefore, determination and concern that the transfer succeed apparently surmount other aspects such as size or longevity.

Training

Needs

Training needs on the respective projects fell into the same general categories regardless of the particular technology involved. The Eastern European workers exhibited greater skills in specific areas than anticipated and did not appear to have extremely low morale problems. But the MNC's did observe that the broad spectrum of competency was often lacking, and that preparation should have been made for supplementing any voids that were observed.

There did appear to be a general lack of positive incentives, combined with sometimes severe retribution for mistakes, that produced an unwillingness and even an inability to accept responsibility or make decisions. Thus, firms might indeed benefit from structuring the transfer so that all decisions were eliminated, or at least so that responsibility for those decisions would not lie with the local workers.

The need for marketing and managerial or supervisory skills was great. The MNCs would have done well to build such training into the plans and cost structure of their transfers, as they often were required to supply such expertise regardless.

Approaches

The training method employed by Honeywell worked very well. Small groups of workers were brought to the United States for training, but these trainees were then responsible for teaching additional employees once back in their own country. This teaching responsibility helped to cement the new information and skills in the original workers, while it spread the processes to others.

Adaption to local conditions was essential; and this included the realization that a normal workday would most likely be impossible if cultural or political events were in the offing. It also included the realization that time off from work would be more frequent in Eastern Europe than in the United States, for reasons not usually acknowledged in this country. Such adaption also included compromises by U.S. employees, such as giving up lunch hours.

Overall, the motto Things Take Time, combined with the admonition "some things will take even more time," might be important to remember during training. It would also be particularly important to build these ideas into areas where quality controls were to be applied. All firms reported that the achievement of certain levels of quality were more difficult to attain than originally anticipated. On the other hand, after acquainting the workers with the reasons for such strict controls, all respective quality levels were achieved eventually.

Prior Experience

Prior experience played an important role in the achievement of quality level, on both the side of the MNC and of the host company or country. Such experience gave the MNC a more realistic level of expectations, in most cases; it gave the host employees a greater awareness of the expectations and quality standards of the MNC.

Prior experience is a two-edged item, however, if past experiences prove detrimental to current training. Prior experience did not seem to be a panacea for all experiences that arose in the course of the transfers. In fact, assumptions based on prior experiences tended to be, at times, misleading. Clark, GT, Bendix, and IH all expected the Eastern European trainees to be familiar with basic manufacturing operations, or portions thereof, and to have a strong orientation to Western quality control standards. As it turned out, Westinghouse and Honeywell, who approached the training issues with knowledge based on past incidents, insisted on strict adherence to each and every instruction from the very beginning, and they succeeded far more quickly. This was despite the fact that in both instances the technology involved was probably more technical and required greater dexterity and quality control. Several firms discovered that <u>un</u>learning or retraining was necessary and also more difficult because the local employees did not realize that their former techniques were not correct. This makes the admonition to follow all instructions exactly even more important, but illustrations regarding the reasons behind such a demand were also seen as most helpful. Company examples along this line included the reasons behind certain safety regulations, or even sanitary or cleanliness standards. On the other hand, unrealistic expectations or standards were never adequately followed, several companies reported.

Communication from one language to another was not a major training problem in any of the transfers. Westinghouse used both languages whenever possible, as well as a system of lights and buzzers designed to facilitate understanding over and above any particular language. Bendix made a point of written confirmation of all verbal messages. Actually, the tone of certain messages was very important. The threatening approach used by the Romanians in some communications with General Tire was detrimental to rapport, but major problems with languages did not appear to jeopardize any of the transfers.

Potential Problems

Disasters

The possibility of major catastrophies jeopardizing a technology transfer always exists, but the likelihood is not that great. The only geographic occurrence affecting any of the transfers in this study was the earthquake in Romania, and while it slowed the project, it did not curtail it. The possibility of disaster certainly encourages the inclusion of the force majeure clause, but it would appear that even the most drastic of events usually will not destroy a transfer in or of itself. Ironically, General Tire noted that the earthquake pulled the two groups together, since both were so adversely affected by the event.

One aspect to consider, however, is the structure of the agreement in the event such an occurrence does take place. Even with application of the force majeure clause, the expenses involved in replacement or repair of equipment or other accompanying problems might easily be made the responsibility of the MNC, unless other procedures are clearly stated.

Delays

Delays occurred in each of the projects, making this aspect perhaps the most inevitable if not undesirable of all project factors. However, the causes of the individual delays varied greatly. An admonition here would be to build in an adequate time estimate that includes possible delays.

Some of the major reasons for the delays encountered in the cases studied here include the following: recessionary economy, which affected both the procurement of and demand for equipment; hard currency shortages, which prevented some materials purchases especially in large quantities; difficulties in equipment installation, tooling, repair, or maintenance; obtaining sufficient raw materials for production, or testing those raw materials to ensure quality control; slow delivery schedules; difficulties in technological developments in either home or host country or both; accidents and a variety of disasters; an underestimation of the time required; problems in achieving required production levels, either related to quality control or other difficulties (this is often a factor stemming from a combination of equipment problems and training needs); and a combination of above factors, which compound individual delays and cause other difficulties such as expiration of warranties.

Some of the delays might be fairly controllable through more careful planning and tighter control over equipment purchase, installation, repair, and usage, and raw materials acquisition; but other delays were more circumstantial and outside the control of the MNC and the local economy. Labor shortages or other labor problems were not a factor in any of the cases involved in this study. In part, this might be one benefit of agreements with a socialist economy, but the possibility of strikes or other stoppages of uncertain duration always remains a factor.

Misinterpretations

Numerous examples were cited regarding differing interpretations of contract terminology. Most of these examples, fortunately, did not involve the technology itself, but rather accompanying facilities. The problem was not one of translation, but interpretation. The term adequate was indicative of one level of accommodation and transportation to members of the MNC and of another level to representatives of the host countries. The suggestion here was to spell out, in detail, the expectations in terms of size, facilities, transportation schedules or availability, and other details, as deemed appropriate. On-site visits would also be most helpful in determining what might be provided or what alternatives were feasible prior to the final contract signing.

Change

The impact of change, particularly on the host country or company, was a factor generally overlooked in these technological transfers. This refers both to a macro, country-wide level and a micro level involving individual workers. Adapting to change was relatively easy on the part of the MNC representatives, as it generally meant little more than adjusting to local cultural conditions. However, the insurgence of high-level technology, particularly in transfers involving massive equipment purchases or radically different production techniques, required major adjustments for many workers from the host nations. International Harvester in particular noted this aspect, pointing out that some of the symptoms that appeared attributable to training problems were really adjustment problems. Firms might do well to include "adapting to change" training in their technological packages.

Development Limits

Joint development projects occurred on several projects. In some instances the development was limited to adapting the technology to the metric system or other local conditions or requirements. In other instances the development took on a far larger magnitude, with completely redesigned products being the final outcome.

The eventuality of such joint developments culminating in additional products seems high enough to warrant some consideration and foresight. Clark Equipment, despite its contract consideration on this subject, ran into a patent conflict over the extent of Clark input into a new product. Most contracts included clauses permitting the MNC to have access to and usage of any products developed from the information from the transfer. Some included fees or royalties to be paid back to the MNC as well as access to the information—a reverse technology flow.

Competition

With high product quality and efficient production among the technology transfer goals, the possibility of eventual competition with the MNC did exist. However, none of the corporations directly involved with these transfers felt any threat at the time of the study. Interestingly, in several instances subsidiaries of the MNC were the ones fearful of possible competition, pointing out the extent of analyses that must be done on the MNC's global operation before such technology transfers are authorized. In the cases here, no reason for concern was felt to exist with regard to the subsidiaries.

Prior to acceptance of the transfers, however, there had been some concern. This was exemplified by the extreme care that was taken in a number of the agreements to delineate the territories that were to be the marketing or sales areas of the respective parties. No infringement thus far had been made on those designated boundaries.

Competition was also not a concern because the products being produced were either not viable products for other areas of the world, including the home territory of the MNC in the United States, or they were in such demand that the concern was to obtain enough products—regardless of the source of the production—to fill marketing needs. Awarding corporate trademarks was a point of leverage here. On one hand, the trademark helped to make the product more immediately competitive; but, on the other hand, unauthorized use of that product would have resulted in the withdrawal of the trademark. As such, the competition threat appeared to lie more in the future, after the expiration of the agreement, when the respective production facilities are on their own and free to market in more areas. At this point, the advancement of the MNC in its own product, combined with its marketing and sales expertise, was expected to be more than adequate to cope with the possibility of competition.

Restructuring

Because the technology transfers involved a period of years, ranging from five to twenty in the examples here, certain corporate changes might be expected. In the instance of IH, the group originally responsible for arranging the transfer was not involved with the operation of it after a major corporate reshuffling. Furthermore, Poland also made some administrative changes on a deparmental level, which also resulted in different personnel involved with the project. Neither of these changes made any appreciable difference on the project, but this was due to the additional care and concern taken by all sides to aid in a smooth transition. In any case, the likelihood of shifts in personnel and administrative arrangements appears high enough to warrant structuring the transfer in such a way that the operations will progress regardless of the personnel involved.

Leverage

Initial Position

A technology transfer without an ongoing, equity basis has often been viewed as undesirable, usually because the ongoing factor helped to eliminate a possible competitor, while the equity provided some basis for control. However, the firms engaging in these nonequity transfers stated that equity of a majority as well as minority level really did not provide sufficient control—particularly when the MNC was in a foreign country. Thus, the MNCs studied here felt that their nonequity stances were no less powerful and, in some instances, were perhaps more potent, because no misconceptions were involved. General Tire stated that the company's position was to consider technology transfers as one type of operation, while the equity/nonequity question was an entirely different set of considerations. The latter related to the financial measures available for such a project, while the former really dealt with the available technology and possible demand.

Firms did make note of the fact that, in the majority of instances, the most powerful position of the MNC in such technology transfers was at the start of the negotiations, when the host desired and/or needed the technology of the MNC. The firms stated that the negotiating sessions were the most viable times in which to utilize the most leverage in gaining desired concessions—financial or otherwise.

A company with a monopolistic hold over certain desired technology is often able to name its price and contract format. This was somewhat descriptive of Honeywell's position. However, if the price is too high, countries unable to afford the technology will settle for another company with similar technology even though it is not the top of the line. In many instances, a company with a monopolistic hold over a certain product or process will refuse to part with this technology, as was the case with Michelin Tires, or Coca-Cola.

Honeywell's decision to supply its technology was partially based on the corporation's assessment of its product's projected life and Poland's potential competitive aspects. Furthermore, despite a successful technology transfer, in that all technology was deemed satisfactorily transferred and all payments were received as scheduled, Honeywell has not become involved with additional transfers to Eastern Europe.

On the other hand, General Tire's willingness to sell its radial tire technology, as well as its base in the United States, allowed Romania to overcome a lack of unique product knowledge.

Sometimes, a firm's reputation and location are more important than its technology. Westinghouse was known for its ability to transfer intricate technology and its location in the United States enhanced its products, which were reputed to be the most advanced and sophisticated in the world. Westinghouse did have superior technology in certain areas, but its negotiating standpoint was enhanced by these other considerations. In addition, a firm whose demands might appear limited by competitors in line for the same project is also able to use this competition to its own advantage. While such situations usually inhibit corporate bargaining, General Tire benefited from knowledge of its competitors in its dealings with Romania. Cognizant that its European competitors did not want to settle for a counterpurchase arrangement or a joint venture, GT was able to exclude these factors from its negotiations as mandatory provisions.

Corporate willingness to work with the Eastern European partner was ranked high—and in some instances, such as GT in Romania, it was seen as more advantageous than the technology itself. Experience in technology transfers, especially the nonequity type, was valuable to both corporations and countries. The corporations benefited from background experiences and a tighter contract format. The countries seemed to rank such experience highly and gave preference to those firms.

Ongoing Leverage

Some portion of leverage or control was essential to the operation throughout the transfer, but each corporation's approach to obtaining such control varied. The idea is essentially the same as at the time of negotiations—have something in reserve that is desired by the other party. A summary of elements that appeared useful to the MNCs in gaining control included updated technology; additional services such as marketing expertise or sales openings; additional technology that might be the basis for another project; the corporate trademark (although this was not always so); product repurchases; or being the sole proprietor of certain technology.

In some instances structuring the agreement to require teamwork was in itself a form of leverage for both sides. A contract that listed the specific responsibilities of both sides, often divided into certain stages, was also seen as virtually essential in many operations. An additional list of crucial duties that were not the responsibility of one party or the other was also seen as beneficial in clearing up misunderstandings and avoiding unexpected expenses.

An element that provided a corporation with control or leverage throughout the transfer process was that of providing technology in stages—always having another part in reserve. Clark Equipment stated that it uses this technique in some transfers if other control elements are lacking. However, it also pointed out that this is not a necessity if the corporation has other items desired by the partner. For example, Clark's marketing connections and expertise were desired by Japan, an element that would keep both parties working together for their common benefit. General Tire had the advantage of possessing technology for off-the-road tire manufacturing in its dealings with Romania. As far as possible then, firms provided the technology over the period of the transfer so that the final information came only as the project was completed. Firms unable to do this had to seek out other forms of leverage that approximated this step-by-step technology transfer or that provided another sought-after element.

Initially the trademark was viewed as a considerable form of leverage; but it emerged as less so if it was not highly visible, or if it became an either/or alternative to financial gain. Payments due to the MNC were the host country's chief form of control, but agreements structured so that delays or other expenses were the host's responsibility tended to have a more rapid completion time than others without this arrangement. Those firms demanding a minimum royalty fee regardless of production also found work delays substantially shortened. However, General Tire's bottom line approach was to maintain the position of being able, at any time, to walk away with a minimum of loss.

Contract Preparation

Negotiations

Mention has already been made of the importance of the negotiating session, in that it is usually the time the MNC has the most leverage. Concessions accorded and provided at this point will usually structure the transfer for its entire duration.

Corporate officials found negotiating sessions with Eastern Europe to be particularly arduous. Sessions were long, arguments were numerous, and extreme care had to be taken to keep prices and concessions within the desired range as quantities or time frames were shifted and revised. Since the contract in a socialist state tended to be the final law and arbitrator, corporations stressed the fact that the contract had to be clear, specific, and detailed. Those areas open to interpretation were seen as detrimental, particularly to the MNC. Personal contact turned out to be of paramount importance. An invitation came because of a firm's reputation, but closure on contracts was made on the basis of personal friendship. The length of the negotiating session was both detrimental and beneficial. Those negotiations that culminated relatively soon usually indicated a higher demand on the part of the host country for the technology. On the other hand, such relative brevity did not permit many strong acquaintanceships to form. In the long run, this personal acquaintanceship had to be generated at one time or another, and some of the longer negotiations paid dividends by having fewer areas that needed to be worked out on the actual projects.

Format

Ambiguity over such terms as adequate has already been mentioned, but the lesson remains in that all general descriptors should be defined. Firms with specific lists of responsibilities for both parties, combined with a list of tasks that were not to be responsibilities, found their transfers made proportionately easier. Furthermore, such a listing provided a form of leverage, especially when combined with certain time segments. Inclusion of contingency measures, such as force majeure clauses or transfer impossibilities, were seen as essential—not because such elements were actually anticipated, but because those measures provided a form of insurance for the MNC and the host party alike.

Facilitating Aids

Political Involvement

Government intervention on either side is usually taken to be an inhibiting factor, but the MNCs in this study found the opposite to be true in many cases. The U.S. government position in the early 1970s (when these transfers were being considered) was most encouraging of trade with Eastern Europe. The respective governments in Eastern Europe were desirous of increasing their exports as well as updating the level of technology in their country. Therefore, the atmosphere on both sides was one of encouragement, and transfers apparently benefited from it accordingly. General Tire noted that high government officials in Romania were willing to lend support and authorization if difficulties arose. Yugoslav officials were able to make decisions regarding the Bendix transfer without even consulting higher authorities.

This is not to say that political difficulties were nonexistent. Several firms, General Tire being a prime example, reported that security harrassment was definitely evident. Visas for host workers coming to the United States for training were sometimes difficult to obtain. Care had to be taken that a certain distance was maintained between the local workers and the MNC officials in some instances.

On the U.S. side, approvals for the transfers were often caught in bureaucratic red tape. Honeywell had to receive clearance for its sensitive sensor technology, and it had to work out the third-country holding arrangement to surmount any political barriers that might arise prior to the conclusion of the project and the transfer of the promised information. On the whole, however, governmental aid made the projects more accepted and operational.

Labor Unions

No labor union objections to any of the transfers were reported, perhaps due to the care that was taken by the individual MNCs to point out that no jobs were really being exported, and that such transfers were actually helping product sales. Individual protests by workers in the United States against dealings with the socialist bloc were noted in one surprising development, but this was not considered a major problem. This appeared to indicate a general revision of thought by individuals in the United States and a greater acceptance of contact with the Soviet bloc than had existed in prior decades.

Personal Contact and Support

Corporate commitment, from the top officials on down, was seen as a prime factor in the eventual success of the projects, in that the MNC would be more willing to work at problems and contend with difficulties that might arise. The necessity of personal acquaintanceship has been mentioned, but it should be reinforced, since this element was virtually vital to transfer operations. As rapport increased, so did communication—with a subsequent decrease in misunderstandings and problems.

Realistic Timing

Realistic timing was particularly crucial for the project. Those projects with unrealistic time estimates were beset by frustrations and feelings of inadequacy that were, in reality, unfounded. The source of such suboptimal time estimates was apparently in factors outside the realm of the project itself, perhaps necessitated by negotiational issues. Regardless, an accurate assessment of the work to be done and the time required to accomplish it should be one of the goals of the project planners.

Transfer Rationale

Financial Compensation

The majority of the MNCs in these transfers opted to accept their project primarily for financial benefits. As all firms reported that payments had been received as scheduled (except when delays had slowed down the entire project), this expectation was apparently fulfilled. However, certain projects proved more expensive than originally anticipated, and this cut into the financial compensation that had been envisioned.

Suggestions to help ensure an adequate financial return on the technology and time invested in these transfers have been summarized as follows:

1. The greater the initial disclosure payments, the better.
2. The sooner the initial disclosure payments, the better.
3. The inclusion of an annual escalator or percentage increase clause is most desirable.
4. The exclusion of penalty or holdback clauses is most desirable.
5. If set prices are required, build in escalator factors or allow for increased prices in the original estimate.
6. Payment of an ongoing, minimum royalty fee, regardless of production, is desirable.
7. Listing responsibilities, including expenses, is desirable, particularly if the MNC's expenses are laid out and the host is responsible for all remaining ones (the contract should state such issues).

The corporations found that delays as well as the more obvious situations such as natural disasters cut into anticipated financial benefits. Increased corporate travel, additional training, or other corporate involvement also proved expensive. Westinghouse found that a highly visible inventory control process reduced some of the waste and, hence, expense, but several firms reported that inventory control was an area that needed additional worker training in itself.

Financial flows apparently had more of a relationship to economic conditions than to project descriptions. Clark found that its agreement with Romania provided more financial gain than its involvement with Poland, although the agreement with Poland was larger and had been in progress for three years longer. This was because the recession hit the Polish project at a crucial time, whereas the Romanian project was less affected by the world economic situation.

Firms found that incidental expenses, in addition to corporate travel, affected their revenues. These expenses included housing, additional transportation measures, and other accompaniments designed to make life in Eastern Europe more homelike for the MNC employees posted there.

On the other hand, several firms reported financial gains over and above those predicted by the agreement estimates due to increased sales of components or other equipment. These firms found the technology transfer particularly rewarding financially, over and above the payments for the technological information itself.

FIGURE 5.1

Analytical Factors of Technology Transfer

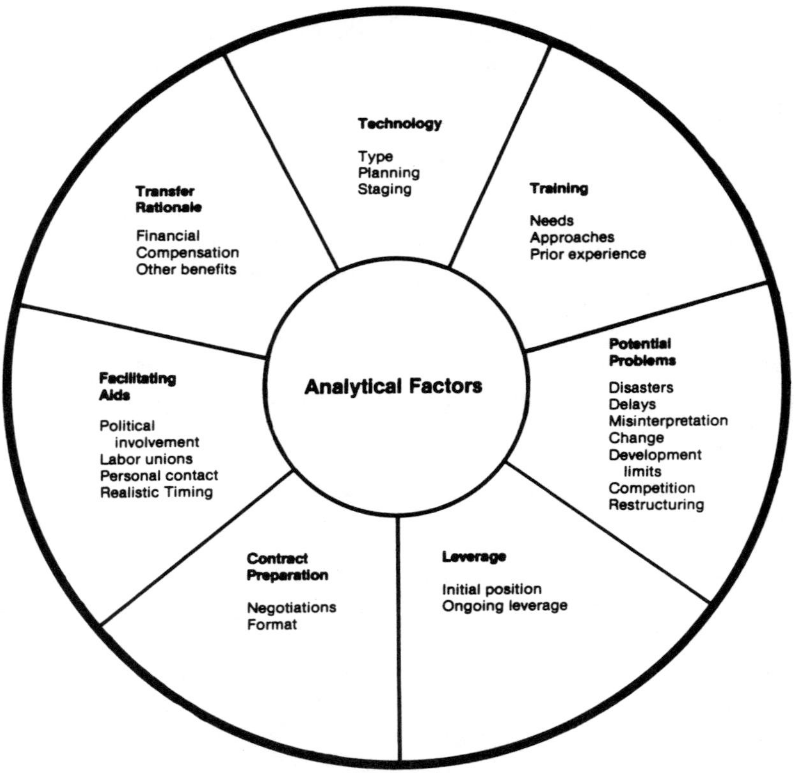

Other Benefits

Financial compensation may have been the initial reason for the technology transfer, but several firms (including International Harvester and Clark Equipment) actually felt that the long-run benefits of the transfers might outweigh the immediate financial gains. Among those benefits cited were alternative product sources and broadened product lines resulting from the joint development activities that were taking place. Additional benefits included access to more information—related particularly to the transfer or even extraneous to it—and a toehold in a market area formerly nonaccessible to the MNC.

The increased market concept, however, was not as fruitful as the MNCs had originally predicted. Two of the six companies did finalize two additional agreements, and many of the firms broadened their initial transfer agreements, but several others, such as Clark and Westinghouse, saw highly desirable contracts awarded to other companies. This aspect actually reduced to an increase in contacts but not necessarily contracts. See Figure 5.1 for analytical results.

CONCLUSIONS

A study of the analytical factors, as drawn from the descriptive ones, provides a type of checklist for the corporation considering a technology transfer to the Eastern European area that should enable the corporation to structure its expectations within more reasonable boundaries than might otherwise have been the case. The analytical factors do not provide easy answers for such a technology transfer but rather present a framework for searching out and analyzing relevant information. See Figure 5.2.

These analytical factors do support the assessment made in Chapter 2 that job exportation and national security issues are issues that, although worthy of consideration, are outside the immediate realm of the multinational corporation in the day-to-day workings of a transfer. These analytical factors also do not conclusively deal with questions relating to competition. The creation of competition was a factor that apparently had not, to date, greatly affected the six MNCs studied here. The firms involved with the transfers did not express any major concerns regarding competition. In fact, the absence of problems regarding marketing territories apparently allayed some prior apprehensions. However, the issue of competition does remain a viable possibility, however remote, for the future.

The analytical factors do provide a summary of the means by which individual firms are structuring their transfers to incorporate some viable leverage or control factors. Such a summary might prove

FIGURE 5.2

Descriptive and Analytical Factors of Technology Transfer

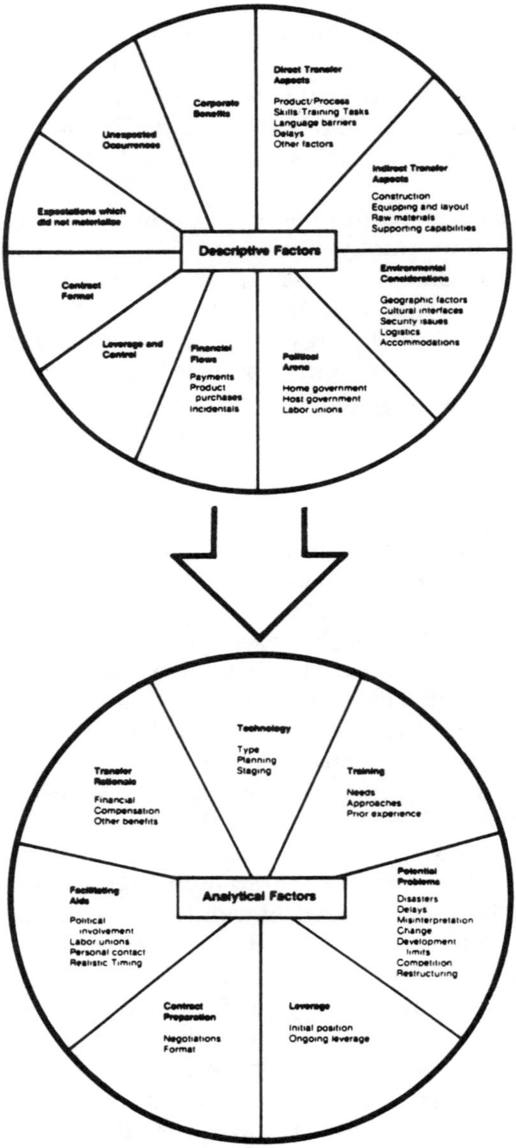

valuable to corporations that have not considered nonequity technology transfers by suggesting leverage potentials. The emphasis on the maintenance of a controlling factor or leverage throughout the transfer cannot be understated—those firms incorporating such an advantage appeared to have the most successful technology transfers, and the balance of power between the two parties provided a climate wherein rapport and mutual understanding was prevalent and satisfactory. See Figure 5.3.

FIGURE 5.3

Possible Remedies for Competition and Control in Technology Transfer

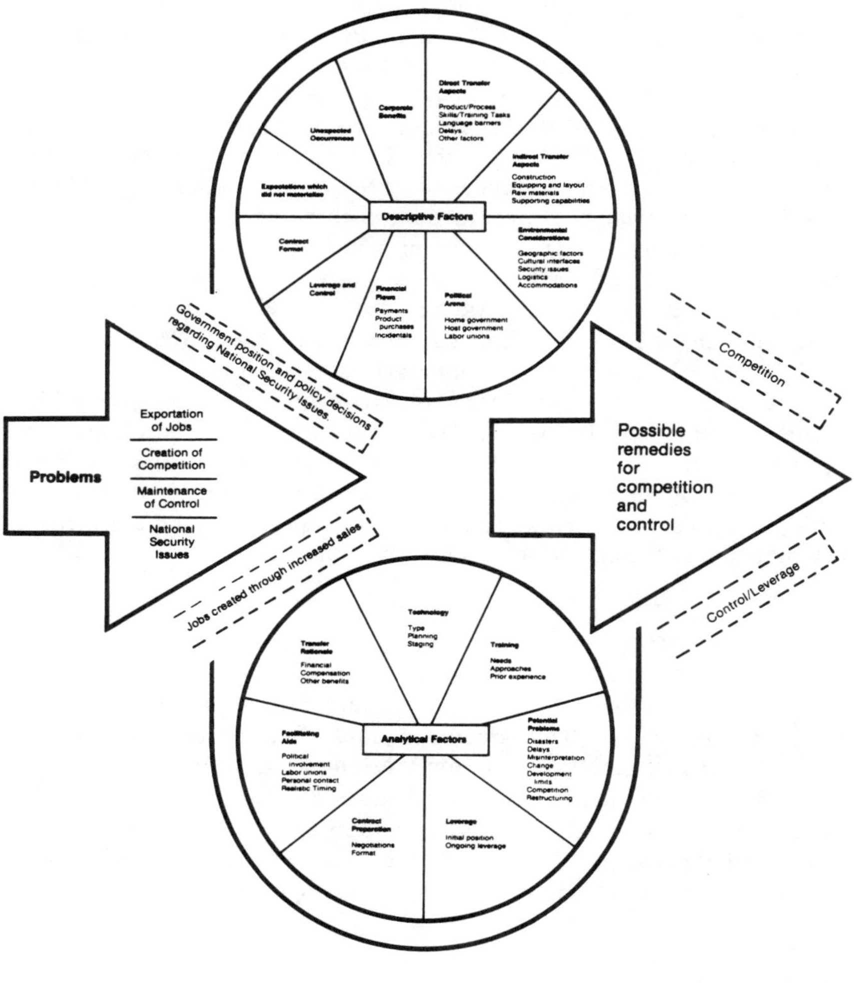

LIMITATIONS OF RESULTS

Some of the information in this research comparing expectations to outcomes came from the contractual terms; however, most of the information originated with individuals whose perceptions must be acknowledged as personal viewpoints. Data from several individuals involved with the same project was used to reduce some of the variations, although it must be realized that communication with several individuals can also distort expectations if their opinions are not reflective of the corporation as a whole.

Some of this research involved agreements that are currently in progress; hence, the results are also in progress and not completed. Variations between expectations and results may change by the time the agreement is completed.

Even within this small sample, the products, countries, and firms have distinct variations—hence, these may affect the outcome more than some of the expected variables.

A relatively small number of situations in one particular region of the world comprised this study. As such, the generalizations derived may not be applicable to other firms, other projects, or other regions. However, the uniqueness of this pilot study is offset by the extension of knowledge it provides on such nontraditional transfers.

This research occurred during a period when relations between the United States and Eastern Europe were highly acceptable. Should political events intervene, however, the favorable governmental climate could fade, and increased contact with this portion of the socialist bloc might be proportionately more difficult.

This study also examined the transfers at a point in time when labor and other conditions were relatively favorable and calm. Upheavals in either the United States or the countries of Eastern Europe would certainly change these results. Such upheavals are seen as being less likely, perhaps, in Eastern Europe or among the MNCs, but they are not impossible or even unrealistic.

FUTURE RESEARCH

Future research in the area of nontraditional transfers is needed in several respects. This study approached such transfers from the viewpoint of the multinational corporation—a similar analysis might be done from the angle of perceived results. Another approach might be to compare the nontraditional transfers in Eastern Europe with those taking place in other socialist regimes outside this bloc, such as the People's Republic of China, or in nonsocialist regions around the world, particularly some of the Third World developing nations. Over a period of time, the entire concept of the nontraditional transfer itself would be an area for study: its increasing or decreasing acceptance and the variations that have been generally made with regard to this form of transfer.

APPENDIX A

TABLE A.1

Countries, Companies, and Product Areas Involved in the Technology Transfers Selected for This Study

Representative Firm in Eastern Europe	Country	Company	Location	Product	Agreements	Time Frame
Bumar Foreign Trade Enterprise	Poland	Clark Equipment Company	Buchanan, Michigan (U.S.)	Cast housing and axles	Technical Assistance Agreement, Trademark Rights, Marketing Agreement (including buy-back)	1972-82 (ten years) plus royalty payments for ten years starting on date of first product sale
Bumar Foreign Trade Enterprise	Poland	International Harvester Company	Chicago, Illinois (U.S.)	Crawler tractors, bulldozers, wheel loaders, and pipelayers	License Agreement, Trademark Rights (with an optional counterpurchase agreement)	1972-82 (ten years initially, now extended to all items to 1990)
Metronex/Mera	Poland	Honeywell	Minneapolis, Minnesota (U.S.)	Class 41 transmitter including silicon sensor	Technical Assistance Agreement (including testing kits); no trademark but Honeywell license acknowledgment	1974-79 (five years) with extension provision
Unitra Electronic Industry Union	Poland	Westinghouse Electric Corporation (Semiconductor Division)	Pittsburgh, Pennsylvania (U.S.)	Rectifiers and thyristors	Technical Assistance Agreement, Patent License Agreement, no trademark but Westinghouse license acknowledgment	1974-79 (five years)
Romania State Enterprise of Foreign Trade (ROMCHIN)	Romania	General Tire International Company	Akron, Ohio (U.S.)	Radial tire technology	Technical Assistance Phase 1 Agreement, Technical Service Contract Phase 2; General Tire trademark available after completion of acceptance tests—license acknowledgment available from start	1973-? (ten-year agreement to run from date of completion of acceptance tests)
Universal Tractor Foreign Trade Company	Romania	Clark Equipment Company	Buchanan, Michigan (U.S.)	Power-shift transmissions and torque converters	Technical Assistance Agreement, Trademark Agreement, Marketing Agreement (including optional buy-back)	1975-85 (ten years for each model, starting on production date for that model and contingent upon royalty payments)
Hungarian Railway Carriage and Machine Works (RABA)	Hungary	International Harvester Company	Chicago, Illinois (U.S.)	Tandem axles, plows and disc harrows, cyclo planters, grain drills	License Agreement, Trademark Agreement, Purchase Agreement	1976-86; 1977-87; 1978-88; 1979-89 (ten-year agreements for each product line)
Varazeinski Tekstini Kombinat (Vartek)	Yugoslavia	The Bendix Corporation	Detroit, Michigan (U.S.)	Organic friction material for disc brake pads	Technical Assistance Agreement	1975-85 (ten years)

APPENDIX B

The questionnaire that follows is the one used as the basis for personal interviews with corporate executives in obtaining the data behind this study of technology transfer. It was used as a guide for the interviews, not as a mail questionnaire. Use of the same structure in all interviews helped to provide a consistent body of information that was fairly uniform throughout all interviews. The points of contrast, therefore, in the collection of the information tended to stem from two sources: variations in the type of technology transfer taking place, or varying corporate reluctance to discuss certain confidential aspects of the transfer.

Questionnaire on Technology Transfer to Eastern Europe

Country: ______________________

Company: ______________________

Interviewee(s): ______________________

Location: ______________________

Date: ______________________

Part 1

Description of Involvement

I. Background

a. General description of corporation as to types of products, number and location of subsidiaries, etc. around the world.

b. Corporation's involvement in Eastern Europe.

1. Prior involvement(s) (excluding agreement to be studied).

2. Future or pending involvement(s)

II. Agreement(s) under study

a. Number (and breakdown) of arrangement(s): (include date(s) (official start, and termination dates)

b. Negotiations leading up to agreement(s): Who instigated, length of time, etc. (Include initial reasons both USA and host party were interested in agreement

c. Types of Involvement: select appropriate one(s):

____1a. TURNKEY (construction of plant):
____1b. TURNKEY (equip plant):
(suggest equipment):

____2. Transfer PRODUCT technology (blueprints, etc):
____3. Transfer PROCESS technology (how to manufacture, etc.):

____4. MANAGEMENT (how to RUN plant):
____5. MANAGEMENT (training others to: (note whether on-site, or elsewhere, or BOTH): operate equipment, manufacture item:

____6. Provide PATENT rights (if so, how long?):

____7. Provide TRADEMARK rights (again how long? [give back?])

____8. Purchase products on buy-back agreement: (details, was this essential?):

____9. Other aspects?

d. Arrangements made with:

1. Country:

2. Group within country:

3. Factory, etc. involved:

e. Responsibilities of USA Corporation under (each) contract:

f. Responsibilities of Eastern Europe Party under (each) contract:

III. Time Frame

What tasks have been involved in the whole operation:

1. Set up:

2. Production:

3. Marketing tasks:

4. Other?

IV. Aspects of Involvement

A. Financial flows:

1. Any monetary payments to the USA corporation?

2. Monetary unit (currency)?

3. Products in exchange:

4. (1) Analyzed by, (2) Within company, (3) Elsewhere in country, (4) Elsewhere in East Europe:

   a. Total units (esp. prior to production):

   b. Parts (sub-assembly--going into production?):

   c. Service pieces (for repairs on prior sales?):

B. Staffing

1. USA Personnel on project

   a. Number of people involved

   b. Visits, or length of time in host country:

   c. Job areas or skills required

2. Costs of personnel (who paid)

   a. Base pay (additional):

   b. Fringes (additional?):

   c. Travel, lodging, etc., abroad?:

3. Problems which have arisen with <u>USA personnel</u>:

   a. Timing of staffing requirements:

   b. Accompanying staffing problems (housing, family discontent, etc.):

   c. Proportion of personnel used in getting production going, versus proportion used in training others:

4. Description of involvement with host personnel:

    a. Number of employees involved:

    b. Types of jobs or tasks to be performed:

    c. Background of nationals:

    d. Additional relevant description:

5. Problems which involved host personnel:

    a. Skills deficiencies?

    b. Morale low?

    c. Quality control lacking?

    d. Language problems?

    e. Other?

C. Protection of Participation

1. Initial protection

    a. Any controls (stated or unstated) in agreement, designed to help USA corporation: (such as holding back certain technology until all else is completed, etc.):

    b. Any controls (stated or unstated) in agreement designed to help host partner (such as does not provide payment until all obligations met, etc.):

    c. What guarantees were provided by USA corporation that all their responsibilities would be completed?

    d. What guarantees were provided by the host partner that all their responsibilities would be completed?

2. On-going protection

   a. Inspection provisions included for:

      1) the USA corporation?

      2) the host partner?

   b. What protection was offered for the:

      1) patents involved?

      2) trademarks involved?

3. External protection: What provisions were included for solving disputes:

   a. Arbitration clauses in contracts (where located, if so)

   b. Force Majeure

   c. Any court involvement?

   d. Other?

V. Problems encountered (other than mentioned above)

   A. Connected with host?

      1. Construction/equipping?

      2. Staffing/training?

      3. Production (supplies, quality, quantity)?

      4. Approvals, red tape holdups?

      5. Others?

   B. With host government?

   C. With the transfer of technology:

      1. Product transfer?

      2. Process transfer?

      3. Management (operating plant)?

      4. Management (training others)?

D. Any problems with unions?

1. In USA?

2. In host country?

E. Any USA governmental opposition?

F. Other problems encountered?

VI. Assessment

A. What has made entire operation <u>easier</u>?

B. What had made entire operation more difficult?

C. <u>Current</u> view of opportunities being provided by this agreement: rank, if possible, for USA corporation. (1) Within country, (2) Within Eastern Europe, (3) Elsewhere

1. Increased exports for:

a. Total assemblies

b. Subassemblies

c. Replacements

2. Toehold in market leading to greater involvement:

3. Fees being paid:

a. To offset prior product development costs:

b. To take advantage of technology before obsolescence occurs?

4. Obtain source of product for supply elsewhere (gaining another manufacturing facility):

5. Obtaining additional supply of raw materials:

6. Means of hard currency from sale of product:

7. Other?

Part 2

Variations or Changes in Involvement

VII. Variations as to:

A. ____ Number or types of contracts comprising the arrangements:

____ Dates of start:

____ Dates of termination: (shortened, lenghtened, other?):

____ Other?: (Revisions in contract format, etc?)

B. Unexpected occurrences connected with NEGOTIATIONS in arranging agreement(s):

1. Sources of unexpected occurrence(s):

2. Reason(s):

3. Adaptation(s):

C. Variations as to types of involvements

1. Necessity to do MORE than anticipated:

2. Necessity to do LESS than anticipated:

3. Other variation(s):

D. Variations in arrangements as to: (more approvals needed, different parties signing, anything?):

1. Country?

2. Group within country?

3. Factory?

E. Variations as to responsibilities of USA corporation?

F. Variations as to responsibilities of Eastern Europe Partner?

VIII. Time Frame variations: What lags, speedups, etc. have occurred? Why?

A. Set up?

B. Production

C. Marketing tasks?

D. Other?

IX. Aspects of Involvement:

A. Financial flows: variations as to amount, timing, or form of payment? (Or, other aspects here?):

1. Monetary payments?

2. Monetary currency?

3. Products?

4. Review what variations occurred from what was expected. (1) Within company, (2) Elsewhere in country, (3) in East Europe:

a. Total unit sales

b. Subassembly sales

c. Service (repair) parts

B. Staffing

1. Variations in USA personnel on project

a. Number of people involved?

b. Visits or length of stay?

c. Job areas or skills involved?

2. Variations in costs of such personnel?

a. Base pay (increase, decrease?)

b. Fringes (increase needed, decrease?)

c. Travel (etc.?)

3. Problems involving USA personnel: what problems were anticipated?

   a. Timing

   b. Accompanying problems?

   c. Operating vrs. training proportion?

   d. Other problems which did not arise but were expected?

4. Description of host involvement: variations from expectations as to:

   a. Number involved?

   b. Tasks required to be performed?

   c. Background of employees?

   d. Other?

5. Anticipated problems involving host personnel:

   a. Skill deficiences?

   b. Low morale?

   c. Low quality control?

   d. Language difficulties?

   e. Other problems anticipated which have not arisen?

C. Protection of participation

1. Initial protection

   a. Have the controls met any anticipated contingencies?

   b. Have the controls been used by the host partner (if they exist?)?

   c. Have the guarantees proved necessary for the USA corporation?

   d. Have the guarantees proved necessary for the host partner?

2. On-going protection

a. Inspection provisions: have they been utilized by:

1) the USA side?

2) the host side?

b. Have any problems arisen regarding: (also, any anticipated problems which have not?)

1) patents?

2) trademarks?

3. External protection: Have any situations utilizing these occurred?

a. Arbitration

b. Force Majeure?

c. Court involvement?

d. Other?

X. Problems encountered; what problems were anticipated which have not occurred?

A. Host problems

1. Construction/equipping/

2. Staffing/training?

3. Production (supplies, quality, quantity)?

4. Approvals, red tape, etc?

5. Other?

B. What problems were anticipated but did NOT appear with host government?

C. What anticipated problems did NOT appear in the transfer?

   1. Product?

   2. Process?

   3. Management (operating plant)?

   4. Management (training others)?

D. Union problems which did not arise?

   1. In USA?

   2. With host?

E. USA governmental opposition which did NOT arise?

F. Other anticipated problems which did not arise?

XI. Hindsight

A. What was expected to make operation easier, but DID NOT?

B. What was expected to make operation more difficult, but DID NOT?

C. View of opportunities when agreement was being considered and first arranged: (rank if possible) (1) Within country, (2) Within Eastern Europe, (3) Elsewhere.

   1. Increased exports for:

      a. Total assemblies

      b. Subassemblies

      c. Replacement parts

   2. Toehold in market leading to greater involvement?

   3. Fees being paid:

      a. Product development cost?

      b. Utilize product before obsolescence sets in

4. Obtain source of supply of product

5. Obtain raw materials

6. Means of hard currency?

7. Other?

BIBLIOGRAPHY

"The Acceleration of International Technology Transfer." Commerce America 29 (March 1976): 9-11.

Acuff, Frank L., and Maurice Villere. "Games Negotiators Play." Business Horizons, February 1976, pp. 70-76.

American Committee on East-West Accord. Common Sense in U.S.-Soviet Relations. April 1978.

Anderson, W. S. "Technology Transfer in Transition." Computers and People 27 (March 1978): 11-12.

Armitage, John A. "The Political Climate and U.S.-Soviet Trade." Columbia Journal of World Business, December 1973, pp. 42-46.

Barbash, Fred. "U.S. Export Licenses Are Linked to Soviet Cooperation." Washington Post, June 26, 1978, p. 45.

"Barriers to Technology Transfer Studied." Aviation Week and Space Technology, August 21, 1972, pp. 12-14.

Basche, James R., Jr. East-West Trade: The Lessons from Experience. New York: Conference Board, 1971.

_____. Evolving Corporate Policy and Organization for East-West Trade. Conference Board Report no. 635. New York: Conference Board, 1974.

Basche, James R., Jr., and Michael C. Duerr. International Transfer of Technology: A Worldwide Survey of Chief Executives. New York: Conference Board, 1975.

Behrman, Jack N. "Multinational Corporations, Transnational Interests and National Sovereignty." Columbia Journal of World Business, March-April 1969, pp. 15-21.

Behrman, Jack N., J. J. Boddewyn, and Ashok Kapoor. International Business-Government Communications. Lexington, Mass.: D. C. Heath, 1975.

Behrman, Jack N., and Harvey W. Wallender. Transfers of Manufacturing Technology within Multinational Enterprises. Cambridge, Mass.: Ballinger, 1976.

Berlin Chamber of Commerce and Industry. Guide to Trade and Cooperation with Eastern Europe: Answers to Practical Questions. Berlin: Chamber of Commerce and Industry, 1974.

Blanning, Robert W., Andrzej W. Kisiel, and Myron Uretsky. "The New Economic System in Poland." Columbia Journal of World Business, December 1973, pp. 113-19.

Bleiberg, Robert M. "Dirty Business: The U.S. Should Ban Technology Transfers to the Soviets." Barron's 58 (July 31, 1978): 7.

Bradley, Gene E. "East-West Trade." Columbia Journal of World Business, December 1973, pp. 39-41.

"The Breakdown of U.S. Innovation." Business Week, February 16, 1976, pp. 56-63.

Brener, Stephen W. "Management Contract." Resort Management, May 1972.

Buchan, David. "Carter Stops Computer Sale to Russians." Financial Times (London), July 19, 1978, p. 1.

Bucy, Fred J. "Technology Transfer: We're Still Selling Them the Rope." Government Executive 10 (September 1978): 38-39.

Business International Corporation. International Licensing: Opportunities and Challenges in Worldwide Technology Management. New York: Business International Corporation, 1977.

Business International S.A. Corporate Strategy, Planning, Organization and Personnel Practices for Eastern Europe. A Group Research Study, summarized results. Geneva: Business International S.A., 1977.

_____. Current Countertrade Policies and Practices in East-West Trade. A Group Research Study. Geneva: Business International S.A., 1976.

_____. Doing Business with Eastern Europe: A Guide for Western Corporations. Geneva: Business International S.A., 1972.

_____. Doing Business with Romania. Geneva: Business International S.A., 1973.

_____. "IH Reorders Priorities to Emphasize Asset Management over Sales." Business International S.A., April 15, 1977, p. 119.

_____. Solving East European Business Problems, vol. 3 (reprinted from Business Eastern Europe). Geneva: Business International S.A., 1977.

Capstick, Richard. "The Perils of Manufacturing Abroad." International Management 33 (March 1978): 43.

Carson, Iain, and Ivana Losco. "Who Are the Capitalists Now in Europe?" Vision, May 1976, pp. 71-75.

Clauser, H. R. "International Technology Transfer—Demands for Change, Foreign Investment and Technology Drain." Research Management, September 1976, pp. 2-3.

Clawson, Robert W., and William F. Kolarik, Jr. "Trade Technology and Soviet R&D: U.S. Corporate Executives' View." Paper presented at the 1976 AAACS Annual Conference (panel title: The State of Soviet Research and Development), October 8, 1976, at Kent State University, Kent, Ohio.

Clutterbuck, David. "Setting Up Shop in the East Bloc." International Management, February 1975, pp. 40-44.

Cogan, Russell E. "Management Development for Nationals Overseas." Personnel, March-April 1973, pp. 24-30.

Coleman, Herbert J. "USAF Probes Foreign Technology Gains." Aviation Week & Space Technology, July 15, 1974, p. 261.

Cooper, Arnold C., and Dan Schendel. "Strategic Responses to Technological Threats." Business Horizons, February 1976, pp. 61-69.

Cordtz, Dan. "Bechtel Thrives on Billion-Dollar Jobs." Fortune, reprinted in Corporate Management in Action, portfolio of case histories, pp. 53-60.

Cotton, Frank E., Jr. "Some Interdisciplinary Problems in Transferring Technology and Management." Management International Review 1 (1973): 59-65.

"Countertrade: Not for Everyone but Worth a Look." Commerce America, June 19, 1978, pp. 2-4.

David, Edward E., Jr. "Technology Export and National Goals." Research Management, January 1974, pp. 12-16.

Diebold, John. "Doing Business in Developing Countries." Financial Executive, December 1973, p. 25.

Dixon, John W. "Technology Transfer—An Essential Export Resource." In The International Essays for Business Decision Makers, edited by Mark B. Winchester, pp. 123-27. Dallas: SMU School of Business, 1976.

Donaghue, Hugh P. "Control Data's Joint Venture in Romania." Columbia Journal of World Business, December 1973, pp. 83-89.

Downey, Arthur T. "East-West Trade: Where Are We Now and Where Will We Be?" Address before the 13th Annual Meeting of the Licensing Executives Society, October 10, 1977, at Palm Beach, Florida.

Ellison, Richard. "An Alternative to Direct Investment Abroad." International Management, June 1976, pp. 25-27.

"Export Controls Split Business." Business Week, May 7, 1979, p. 36.

"Exports Are Only One Way to Sell Abroad." Industry Week, August 18, 1975, pp. 12-13.

Fahey, Thomas. "Trading and Investing in Eastern Europe: A Businessman's Perspective." In China, the U.S.S.R. and Eastern Europe: A U.S. Trade Perspective, edited by John K. Ryans, Jr., Anant R. Negandhi, and James C. Baker, pp. 132-42. Kent, Ohio: Kent State University Press, 1974.

Fforde, J. S. An International Trade in Managerial Skills. Oxford: Basil Blackwell, 1957.

Ford, David, and Chris Ryan. "The Marketing of Technology." European Journal of Marketing (United Kingdom) 11 (1977): 369-82.

Frank, Isaiah. "The Export Answer." Columbia Journal of World Business, May-June 1968, pp. 17-22.

"French Exports See Future Competition." Business Week, May 23, 1977, p. 34.

Gabriel, Peter P. "Adaptation: The Name of the MNC's Game." Columbia Journal of World Business, November-December 1972, pp. 7-14.

_____. "From Capital Mobilizer to Management Seller." Columbia Journal of World Business, March-April 1967, pp. 7-16.

_____. The International Transfer of Corporate Skills: Management Contracts in Less Developed Countries. Boston: Harvard University Press, 1967.

_____. "The Investment in the LDC: Asset with a Fixed Maturity." Columbia Journal of World Business, Summer 1966, pp. 109-19.

_____. "More Channels Needed to Transmit Management Know-how." Columbia Journal of World Business, January-February 1969, pp. 33-40.

_____. "MNC's in the Third World: Is Conflict Unavoidable?" Harvard Business Review, July-August 1972, pp. 93-102.

Gall, Peter, and Lewis H. Young. "How Russia Would Spur Trade with the West." Business Week, December 27, 1976, p. 34.

Gibson, W. David, and John Price. "A New Ball Game for Multinationals in the Offing." Chemical Week, December 18, 1974, p. 29.

Giffen, James H. "Developing a Market Program for the U.S.S.R." Columbia Journal of World Business, December 1973, pp. 61-68.

Green, Donald W., and Gerbert S. Levine. "Implications of Technology Transfers for the U.S.S.R." Paper presented at the Annual Meeting of the Eastern Economic Association, April 1976, at Bloomsburg, Pa.

Guffey, Al. "Some Experiences of Occidental Petroleum in Negotiating Cooperation Agreements in the U.S.S.R." International Lawyer vol. 1, pp. 41-53.

Hanson, Walter E. "How the Government Can Smooth International Trade." Commercial and Financial Chronicle, February 3, 1975, p. 9.

Harvey, M. G., and R. A. Kerin. "Multinational Corporations Versus Organized Labor: Divergent Views on Domestic Unemployment." California Management Review 18 (Spring 1976): 5-13.

Hawkins, R. G., and B. Finn. "Regulation of Multinational Firms' Foreign Activities: Home Country Policies and Concerns." Journal of Contemporary Business 6 (Autumn 1977): 7-30.

Hayden, Eric W. Technology Transfer to East Europe: U.S. Corporate Experience. New York: Praeger, 1976.

_____. "Technology Transfer to the Soviet Bloc." Michigan State University Business Topics, Winter 1976, pp. 11-23.

_____. "Transferring Technology to the Soviet Bloc: U.S. Corporate Experiences." Research Management, September 1976, pp. 17-23.

Hayden, Eric W., and Henry R. Nau. "East-West Technology Transfer: Theoretical Models and Practical Experiences." Columbia Journal of World Business, Fall 1975, pp. 70-82.

Hertzfeld, Jeffrey M. "Setting Up Shop in Moscow." Harvard Business Review, September-October 1974, pp. 137-42.

Hesseling, Pjotr. "Studies in Cross-cultural Organization." Columbia Journal of World Business, December 1973, pp. 120-34.

Hewett, Edward A. "The Economics of East European Technology Imports from the West." American Economic Association, May 1975, pp. 377-88.

Holt, John B. "New Roles for Western Multinationals in Eastern Europe." Columbia Journal of World Business, Fall 1973, pp. 131-39.

Hosoya, Chihiro, Henry Own, and Andrew Shonfield. Collaborating with Communist Countries in Managing Global Problems: An

Examination of the Options. New York: The Trilateral Commission, 1977.

Huhs, John I. "Developing Trade with the Soviet Union." Columbia Journal of World Business, Fall 1973, pp. 116-30.

Hull, Cordell W. "Practical Considerations of Middle East Service Agreements." Intertrade Journal, June-July 1976, p. 24.

Huntington, Samuel P. "Trade, Technology and Leverage: Economic Diplomacy." Foreign Policy 32 (Fall 1978): 66-79.

International Chamber of Commerce. "The Future of East-West Trade." March 1978.

"International Transfer of Technology: Sources of Conflict." Paper presented by the U.S. Delegation to the Third Quadrilateral Businessmen's Conference on Economic Interdependence, London, England. U.S. Chamber of Commerce, Washington, D.C.

Janssen, Richard J. "Debt Threat: Soviet Bloc Borrowings from the West Surge, Amid Mystery, Fears." Wall Street Journal, February 22, 1977, p. 1.

Jones, David. "The 'Extra-Costs' in Europe's Biggest Synthetic Fiber Complex at Mogilev, U.S.S.R." Worldwide R&I, May-June 1973, pp. 30-35.

Kapoor, Ashok. "MNC Negotiations: Characteristics and Planning Implications." Columbia Journal of World Business, Winter 1974, pp. 121-30.

Kapoor Ashok, and J. J. Boddewyn. International Business-Government Relations: U.S. Corporate Experience in Asia and Western Europe. New York: American Management Association, 1973.

Kendall, Donald M. "Present and Future Prospects for East-West Trade." In The International Essays for Business Decision Makers, edited by Mark B. Winchester, pp. 1-7. Dallas: SMU School of Business, 1976.

Keohane, Robert O. "Not Innocents Abroad: American Multinational Corporations and the United States Government." Comparative Politics 8 (January 1976): 307-20.

Kirchner, Jake. "Study Sees No Cause to Bar Data Processing Exports to East." Computerworld 12 (October 16, 1978): 9.

Kiser, John, III. "Is the West's Technology 'Lead' an Illusion?" Government Executive 11 (March 1979): 39-40.

_____. "Report on the Potential for Technology Transfer from the Soviet Union to the United States." Study prepared for the Department of State and the National Science Foundation under Contract No. 1722-620217, July 1977.

Klass, Philip J. "Technology Export Guidelines Pondered." Aviation Week and Space Technology, July 29, 1974, pp. 18-19.

Kovaly, Kenneth A. "Market for Technology Keeps Getting Hotter." Industry Week, July 1, 1974, pp. 36-39.

Kozicharow, Eugene. "Control Revisions Urged to Expand U.S. Exports." Aviation Week, April 5, 1976, pp. 18-19.

Lamm, Lewis L. Address to the Foreign Trade Convention, November 7, 1975. Unpublished paper.

Lang, Eugene M. "Technology Transfer: A Quick Route to Foreign Markets." Management Review, September 1973, p. 62.

Lange, Irene, and James F. Elliot. "U.S. Role in East-West Trade: An Appraisal." Journal of International Business Studies 8 (Fall-Winter 1977): 5-26.

"Latin America Opens the Door to Foreign Investment Again." Business Week, no. 2444 (August 9, 1976), pp. 35-36.

Lauter, G. Peter. "East-West Trade Organizational Structures and Problems of U.S. Multinational Corporations." Paper presented to the 3rd U.S.-Romanian Seminar on Trade and Economic Cooperation, July 17-22, 1978, at Bucharest, Romania.

Lauter, G. Peter, and Paul M. Dickie. "Multinational Corporations in Eastern European Socialist Economies." Journal of Marketing, October 1975, pp. 40-46.

Lecht, Charles P. "Tsunami." Computerworld 12 (June 12, 1978): 49.

McMenamin, Robert J. "Western Technology and the Soviets in the 1980's." NATO Colloquium, Brussels, 1978.

McQuade, Lawrence C., Chairman. East-West Trade: Managing Encounter and Accommodation. The Atlantic Council Committee on East-West Trade. Boulder, Colo.: Westview Press, 1977.

"Management Contracts Grow More Common, But Some Firms Balk." Wall Street Journal, Labor Letter, August 24, 1976, p. 1.

Mann, Clarence J. "The Technology of Mass Merchandising: Its Contribution to Economic Development." Fund for Multinational Management Education and the International Chamber of Commerce (Seminar on International Issues in Transfer of Technology, September 21-23, 1977, Paris, France), paper presentation.

Mansfield, Edwin. "East-West Technological Transfer Issues and Problems, International Technology Transfer: Forms, Resource Requirements, and Policies." American Economic Association 65 (May 1975): 372-76.

_____. "Economic Impact of International Technology Transfer." Research Management, January 1974, pp. 7-11.

"Many Problems Hinder Technology Transfer." Industry Week, June 26, 1972, pp. 20-22.

March, Carl, ed. Common Sense in U.S.-Soviet Relations. Washington, D.C.: American Committee on East-West Accord, 1978.

Marer, Paul. Soviet and East European Foreign Trade, 1946-1969. Bloomington: Indiana University Press, 1972.

Marer, Paul, and Joseph C. Miller. "U.S. Participation in East-West Industrial Cooperation Agreements." Journal of International Business Studies 8 (Fall-Winter 1977): 17-29.

Mason, R. Hal. "The Selection of Technology: A Continuing Dilemma." Columbia Journal of World Business, Summer 1974, pp. 29-34.

Mayer, Charles S. "Marketing in Eastern European Socialist Countries." University of Michigan Business Review, January 1976, pp. 16-21.

Meyer, Herbert E. "Those Worrisome Technology Exports." Fortune 97 (May 22, 1978): 106.

_____. "Why That Soviet Buying Spree Won't Last." Fortune, January 1975, pp. 94-138.

_____. "Why Business Has a Stake in Keeping Sovietology Alive." Fortune, September 1975, pp. 156-62.

Milosh, E. J. "Imaginative Marketing in Eastern Europe." Columbia Journal of World Business, December 1973, pp. 69-72.

Morgan, John P. "The Financial Aspects of East-West Trade." Columbia Journal of World Business, December 1973, pp. 51-56.

Mountain, Maurice J. "The Continuing Complexities of Technology Transfer." Government Executive 2 (January 1979): 42-52.

Moyer, Reed. "Foreign Investment Grows, Changes, Prospers." Columbia Journal of World Business, March-April 1968, pp. 59-65.

"Multinationals Find the Going Rougher." Business Week, July 14, 1975, pp. 64-66.

Newman, William H. "Is Management Exportable?" Columbia Journal of World Business, January-February 1970, pp. 7-18.

Norris, William C. "High Technology Trade with the Communists." Datamation 24 (January 1978): 99-103.

Nye, Joseph S., Jr. "Multinational Corporations in World Politics." Foreign Affairs 53 (October 1974): 153-75.

Oliver, John B., and Elliott J. Weiss. "Is Selling Technology to the Soviets Dangerous?" Harvard Business Review, January-February 1975, p. 18.

Organization for Economic Cooperation and Development. "East-West Trade: Trends in East-West Trade," Working Paper of the Trade Committee. Paris: OECD, 1977.

Patolichev, Nikolai S. "How Russia Would Spur Trade with the West." Business Week, December 2, 1976, pp. 34-36.

Perlmutter, Howard V., Franklin Root, and Leo V. Plante. "Responses of U.S.-Based MNC's to Alternative Public Policy Futures." Columbia Journal of World Business, Fall 1973, pp. 78-86.

Polish Machine Industry Offers 5 (1979): 4.

Prekopa, G. "Technology Transfer to Eastern Europe: General Tire's Experience in Romania." Panel presentation at the 75th Annual Meeting of the National Machine Tool Builders Associations, October 29, 1976, p. 6.

"The Profitable Lure of Exporting Innovation." Business Week, no. 2586 (Industrial Edition) (May 21, 1979), pp. 40J, 40N.

Publicom. Your Commercial Partners in Romania. Publicom, 1977.

Rath, H. J. "Management Techniques in International Technology Transfers." Journal of Business Communication 13 (Winter 1976): 25.

"Regulatory Maze Seen as Problem for U.S. Firms." Computerworld 12 (January 16, 1978): 71.

Rendell, Robert S. "Eximbank Assistance to Trade and Investment in Eastern Europe." International Lawyer 10 (1): 79-84.

Richman, Barry M. "Ideology and Management: The Soviet Evolution." Columbia Journal of World Business, March-April 1971, pp. 62-72.

_____. "Multinational Corporations and Communist Nations." Management International Review 16 (1976): 9-22.

"Rise in Foreign Controls on U.S. Concerns That Operate Abroad Said to Be Problem." Wall Street Journal, November 19, 1976.

Robock, Stefan H. "Political Risk: Identification and Assessment." Columbia Journal of World Business, July-August 1971.

Root, W. A. "Controls on West-to-East Technology Transfer." Defense Systems Management Review 2 (Winter 1979): 46.

Rose, Sanford. "Multinational Corporations in a Tough New World." Fortune, August 1973, p. 52.

Ryans, John K., Jr. "China: An Old Market Becomes the Newest Trade Opportunity." In China, the U.S.S.R. and Eastern Europe: A U.S. Trade Perspective, edited by John K. Ryans, Jr., Anant R. Negandhi, and James C. Baker, p. 15. Kent, Ohio: Kent State University Press, 1974.

"Russia and the West: Technology with Strings?" Economist, April 17, 1976, pp. 50-51.

Salusinszky, Istvan. "Financing Investment in the Eastern European Countries." International Lawyer 10 (1): 65-77.

Sauer, Walter C. "Eximbank Credits Back East-West Trade." Columbia Journal of World Business, December 1973, pp. 57-60.

Schmidt, Robert D. "Technology Transfer: More Rope or Which Rope?" Government Executive 10 (December 1978): 18-21.

Schukin, George S. "The Soviet Position on Trade with the United States." Columbia Journal of World Business, December 1973, pp. 47-50.

"A Sense of Humor Helps." Wall Street Journal, Business Bulletins, October 10, 1978, p. 1.

"Should Communists Buy U.S. Technology?" Industry Week, December 1, 1975, pp. 3-4.

Siligman, Daniel. "Communist Ideology and Soviet Power." Fortune, January 1976, pp. 112-92.

Small, Brian W. "How to Transfer Technology." Management Today, September 1977, pp. 19-26.

Smith, Donald. "Why Companies Balk at Technology Transfers." Columbia Journal of World Business, May-June 1967, pp. 45-53.

"Soviets Bring in Locals to Tell Approaches to Export Markets." Industrial Marketing, January 1976, pp. 26-27.

"Soviet Trade: The Disappointments Continue." Business Week, December 6, 1976, pp. 100-1.

"Soviet Union—Bendix Breaks Ground in Trade with Russia." Business Week, January 31, 1977, p. 49.

Spiers, Joseph N. "More Technology Likely in Imports from U.S.S.R." Industry Week, November 3, 1975, p. 37.

_____. "Technology Exchanges: A Route to Russian Markets?" Industry Week, April 14, 1975, pp. 7-8.

"Spotlight on Technology Export." Aviation Week & Space Technology, July 1, 1974, p. 7.

Stanford Research Institute. Compensation Agreements in the U.S.-U.S.S.R. Economic Relations. Stanford Research Institute, Technical Note SSC-TN-5552-1. Stanford, Calif.: SRI International, October 1978.

Starr, Robert I. "The Potential for Joint Ventures in Eastern Europe." International Lawyer 10 (1): 35-39.

Steele, Lowell W. "Barriers to International Technology Transfer." Research Management, January 1974, pp. 17-21.

Szumski, Jerzy, and Janusz Piotrowski. "Bumar (Poland)-International Harvester (U.S.A.) Cooperation Is Now over Five Years Old and Keeps Going." Office of Polish Commercial Counsel, Chicago, May 1978.

"Technology Exchange with the U.S.S.R." Research Management, July 1974, pp. 7-20.

"Technology Promotion Policies of Five Nations Reported in New Study." Commerce Today (Office of the Assistant Secretary, Science and Technology), February 5, 1973, pp. 9-12.

"Technology Transfer—Is the U.S. Giving Away the Farm?" Government Executive, V. E., 9 (July 1977): 25.

Teplin, Mary Frances. "U.S. International Transactions in Royalties and Fees: Their Relationship to the Transfer of Technology." Survey of Current Business, December 1973, pp. 14-18.

Ulman, Neil. "Russia Finds Building of Biggest Truck Plant Can Be Big Headache." Wall Street Journal, June 23, 1976, p. 1.

"U.S. Companies Turning to Overseas Licensing of Industrial Know-how." Commerce Today (Bureau of International Commerce), June 12, 1972, pp. 7-10.

U.S., Congress. American Role in East-West Trade. Prepared for the Committee on Commerce, 95th Cong., 1st sess., February 1977.

U.S., Congress. General Accounting Office. The Government's Role in East-West Trade—Problems and Issues. GAO Report no. ID-76-13A, February 1976.

U.S., Congress, House, Committee on Science and Astronautics. The Technology Balance, U.S.-U.S.S.R. Advanced Technology Transfer. Hearings before the Subcommittee on International Cooperation in Science and Space, 93d Cong., 1st and 2d sess., 1973.

U.S., Congress, House, Committee on Science and Technology. Background Material on U.S.-U.S.S.R. Cooperative Agreements in Science and Technology. H. Document 613360, 94th Cong., 1st sess., 1975.

U.S., Congress, Library of Congress, Congressional Research Service. East-West Trade and Commercial Relations during 1977. Ronda Bresnick, Washington, D.C., 1977.

_____. A Reassessment of U.S. Export Licensing in East-West Commercial Relations. Prepared for the House International Affairs Committee by John P. Hardt. Washington, D.C., 1976.

_____. Western Technology and Economic Performance in the Eastern Countries. Prepared by John P. Hardt and George D. Holliday, for HC 56 Europe, East, House International Relations Committee Hearings on Export Administration, June 1976.

U.S., Congress, Senate. Committee on Finance. Background Materials Relating to the United States-Romanian Trade and the Extension of the President's Authority to Waive Section 402 of the Trade Act of 1974. 94th Cong., 2d sess., 1976.

_____. Summary and Analysis of H.R. 10710—The Trade Reform Act of 1973. 93d Cong., 2d sess., 1974.

U.S., Department of Commerce. The United States Role in East-West Trade: Problems and Prospects. An Assessment by Rogers Morton, August 1975.

U.S., Department of Commerce, Bureau of East-West Trade. Counter-Trade Practices in Eastern Europe. April 1977.

_____. East-West Trade Financing: An Introductory Guide. September 1976.

_____. East-West Trade Update: A Commercial Fact Sheet for U.S. Business. December 1977.

_____. Impact of Soviet Ministry Management Practices on the Assimilation of Imported Process Technology. Prepared by John P. Young, April 7, 1978.

_____. An Introduction to the Bureau of East-West Trade. Washington, D.C., 1974.

_____. Issues of Technology Transfer in East-West Industrial Cooperation: The U.S. Perspective. Prepared by John Holt, Paul Marer, and Joseph Miller, July 1975.

_____. Joint Venture Agreements in Romania: Background for Implementation. Prepared by Pompiliu Verzariu, Jr., and Jay A. Blurgess, June 1977.

_____. "Major East-West Trade Issues." Washington, D.C., mimeographed.

_____. Preliminary Assessment of Raymond Vernon's and Marshall Goldman's U.S. Policies in the Sale of Technology to the U.S.S.R. Prepared by Thomas A. Wolf, Washington, D.C., 1974.

U.S., Department of Commerce, Bureau of East-West Trade, Office of East-West Policy and Planning. Quantification of Western Exports of High Technology Products to the Communist Countries. Prepared by John P. Young, Washington, D.C., 1978.

_____. Report on East-West Technology Transfer Research and Policy Development: First Quarter 1978. Washington, D.C., 1978.

_____. U.S. Policies in the Sale of Technology to the U.S.S.R. Prepared by Raymond Vernon and Marshall Goldman, Washington, D.C., 1974.

_____. U.S.-U.S.S.R. Trade and the Whipsaw Controversy. August 1977.

U.S., Department of Commerce, Industry and Trade Administration. East-West Countertrade Practices: An Introductory Guide for Business. Prepared by Pompiliu Verzariu, Scott Bozek, and JeNelle Matheson, August 1978.

_____. Working Onsite in Eastern Europe and the Soviet Union. June 1977.

U.S., Department of Commerce, Industry and Trade Administration, Office of East-West Trade Policy and Planning. Selected Trade and Economic Data of the Centrally-Planned Economies. June 1979.

_____. U.S. Trade Status with Communist Countries. November 15, 1979.

U.S., Department of Labor. International Transfer of Industrial Technology by U.S. Firms and Their Implications for the U.S. Economy. Prepared by Jack Baranson, December 1976.

U.S., Department of State. Transfer of U.S. Technology to the Soviet Union: Impact of U.S. Commercial Interest. Prepared by Herbert S. Levine, M. Mark Earle, Jr., Charles H. Movit, and Anne R. Lieberman. Stanford, Calif.: Strategic Studies Center, Stanford Research Institute, February 1976.

"U.S. Must Improve Rules for Trade with Eastern Europe." Marketing News 10 (April 22, 1977): 11-12.

Van Dam, Andre. "Marketing in the New International Economic Order." Journal of Marketing, January 1977, pp. 19-23.

Vernon, Raymond. Storm Over the Multinationals: The Real Issues. Cambridge, Mass.: Harvard University Press, 1977.

Walczak, Donald A. "A System for Transfer of Manufacturing Capability." Unpublished. Westinghouse Semiconductor Division, Pittsburgh, 1978.

Wallender, Harvey W. "Managing and Evaluating Technology Transfer in Manufacturing Industries." UN Economic and Social Council (Seminar on the Management of the Transfer of Technology within Industrial Cooperation, Geneva, July 14-17, 1975), Sc Tech/Sem. 3/R.16, June 13, 1975.

Wasowski, Stanislaw, ed. East-West Trade and the Technology Gap: A Political and Economic Appraisal. New York: Praeger, 1970.

Weigand, Robert E. "International Trade without Money." Harvard Business Review, November-December 1977, p. 28.

_____. "Selling Soviet Buyers." Michigan State University Business Topics, Spring 1976, pp. 15-21.

"The West Gives Everything Away." U.S. News and World Report, March 15, 1976, p. 23.

"Westinghouse Turnkey Contract Typifies Poland's Need for Top Flight Technology." Business International 20 (December 14, 1973): 393-94.

White, Edward P., ed. United States/Eastern Europe: Technology and Patents, Sale and/or Licensing. Licensing Executives Society, 1977.

Wiles, Peter. "World Technological Leadership." Lloyds Bank Review (United Kingdom), no. 127, January 1978, pp. 15-27.

Wilczynski, J. The Multinationals and East-West Relations: Towards Transideological Collaboration. Boulder, Colo.: Westview Press, 1976.

Wilson, T. "To Maintain the Aerospace Industry's Economic Health." Vital Speeches of the Day 41 (September 1, 1975): 687.

Winpisinger, W. W. "The Case against Exporting U.S. Technology." Research Management 21 (March 1978): 20.

Winter, David. "Protecting East-West Trade Contracts and Operations." Columbia Journal of World Business, December 1973, pp. 73-77.

Wolff, Howard. "Polish Plant Ready to Roll." Electronics, October 14, 1976, p. 75.

Yergin, Daniel. "Politics and Soviet-American Trade: The Three Questions." Foreign Affairs 55 (April 1977): 517-38.

"You Don't Sell Soviets—They Have a Shopping List." Industrial Marketing, July 1976, p. 6.

Zoubek, Jan. Prospects of East-West Joint Ventures. East-West Research and Advisory, rue Hobbema, Research Report no. 7, Brussels, May 1974.

Zurawicki, Leon. "The Cooperation of the Socialist State with the MNC's." Columbia Journal of World Business, Spring 1975, pp. 109-15.

Personal Corporate Interviews

Clark Equipment Company, Buchanan, Michigan: December 1977, March 1978, May 1978; Jackson, Michigan: June 1978.

Honeywell, Minneapolis, Minnesota: April 1978, July 1978, August 1979; Fort Washington, Pennsylvania: June 1979.

International Harvester, Chicago, Illinois: December 1977, June 1978, July 1978, November 1979; Schaumburg, Illinois: July 1978.

General Tire International Company, Akron, Ohio: July 1978, June 1979.

The Bendix Corporation, Southfield, Illinois: December 1977, March 1978, November 1978, June 1979, December 1979.

Westinghouse Electric Corporation, Semiconductor Division, Pittsburgh, Pennsylvania: July 1978, June 1979, October 1979.

ABOUT THE AUTHOR

MARILYN L. LIEBRENZ is an Assistant Professor of Business Administration in the School of Government and Business Administration at The George Washington University. Her teaching responsibilities include marketing, marketing management, international marketing, sales management, international business, and multinational corporations in the world economy.

Liebrenz holds a Ph.D. in business administration from the Graduate School of Business Administration at Michigan State University, an M.A. in communications from Michigan State University, and an undergraduate degree in education from Wheaton College.